SUN ROOMS

SUN ROOMS

Create A Beautiful Enclosed Glass Extension for Your Home

by Cheri Rae Wolpert

HPBooks
a division of
PRICE STERN SLOAN
Los Angeles

A FRIEDMAN GROUP BOOK

Published by HP Books,
a division of
Price Stern Sloan, Inc.
360 North La Cienega Boulevard
Los Angeles, California 90048

Library of Congress Cataloging-in-Publication Data

Wolpert, Cheri Rae.
    Sun rooms

    Includes index.
    1. Sun rooms.   2. Greenhouse gardening.   3. Solar
heating.   I. Title.
    TH3000.S85W65   1989   643'.5   88-21239
    ISBN 0-89586-738-9   ISBN 0-89586-739-7 (pbk.)

SUN ROOMS:
Create a Beautiful Enclosed Glass Extention for Your Home
was prepared and produced by
Michael Friedman Publishing Group, Inc.
15 West 26th Street
New York, New York 10010

Editor: Sharon Kalman
Designer: Fran Waldmann
Art Director: Mary Moriarty
Photography Editor: Christopher Bain
Illustrator: Ken Spengler
Production Coordinator: Karen L. Greenberg

Typeset by BPE Graphics, Inc.
Color separations by South Seas International Press Ltd.
Printed and bound in Hong Kong by Leefung-Asco Printers Ltd.

Acknowledgments

The completion of this book was no less daunting a task than the construction of a sun space. But the task was made easier and more enjoyable thanks to the expertise, good wishes, and generous spirits of so many who helped. That help ranged from technical tips to late-night phone calls offering encouragement and support. To all who shared of themselves, I am grateful, especially to industry experts Mike Libby and Wally Ford, and to friends and family: Pete Chaffey, Gail Foley, Bob Howells, John Lehrer, John McKinney, Alice Mendoza, Cindy Robinson, and Ray and Mary Wolpert. And a special thanks to editor Sharon Kalman whose patience and careful attention to detail were highly appreciated.

Courtesy Lindal Cedar Homes, Inc.

CONTENTS

Introduction
Page 9

Introduction

"Cocooning," a term recently reintroduced into our vocabulary, reflects the desire of many of us to spend more time at home, taking a breather from the frantic pace of the outside world. We have begun to seek refuge from congested highways, long lines, poor service and mediocre food at expensive restaurants, and crowded vacation spots. We've learned that the pleasures of home can be every bit as rewarding as any found miles away.

A home can be more than a refuge or a retreat; it can become almost a resort. Let your mind drift for a few moments while you imagine your favorite vacation spot. Is it an idyllic beach, complete with swaying palm trees, warm sunshine, and cool breezes? Or is it a mountain setting, near a lovely stream that trickles down among the ferns? Perhaps it's a warm desert, full of cactus flowers with a hot spring close by. Whatever your dreams of a great getaway may be, you can incorporate those dreams into your home in a very special room—a sun space.

What exactly are sun spaces? Although they are known by a number of names—solarium, solar greenhouse, solar room—they all combine the simple elegance of light, glass, attractive frames, and natural materials that bring the outdoors inside. They are a throwback, really, to the ornate conservatories and greenhouses found in Victorian England. The Victorians, never known for their hedonistic tendencies, used their glass rooms primarily to pursue their passion for gardening. Today, while this remains one of the primary attractions of building a sun space, these rooms can be designed to accommodate a wide range of activities—from working to working out—satisfying everything from your most practical needs to your most indulgent fantasies.

The original sun spaces were necessarily limited in structure and design to the types of building materials available at the time; they were built of wood, iron, and flat panes of glass. Today, the technology borrowed from space research has led to advances in

glazing design, including reflective coatings and tints, along with breakthroughs in super-efficient caulks and other sealants. Additionally, the use of lightweight, extruded aluminum and laminated wood as framing materials has permitted the design of strong curved and arched beams. The blending of high-tech materials, a little imagination, and a new trend in lifestyles has created a real boom in the building of sun spaces. More than 300,000 sun spaces have been built in the last decade, adding up to a $1.5 billion industry.

Sun spaces may be constructed to provide solar heat for the home; they may enclose a hot tub, spa, or even a full-size pool; they may be used as an office, an exercise room, or a family relaxation area. Whatever the function of a sun space, you can be sure that it will add charm and drama to your home—and, not incidentally, a sun space will also substantially increase the value of your home. The cost of adding a sun space is recovered at a rate of 90 to 100 percent, considerably higher than the 55 to 60 percent for a typical room addition.

If you're exhausted by the end of each day, after fighting traffic and dealing with the ever-increasing stresses of daily life; if you want to spend more time enjoying your life, your family, and your home; then you might want to consider "cocooning"—in a sun space designed just for you.

A sun space is the perfect place to unwind; it's the ultimate in "cocooning."

Courtesy Jardin Inc.

CHAPTER 1

Planning Your Sun Space

Before designing and building your sun space, it is essential that you decide for yourself what you want from it. Although it seems a simple task, it's the first and most important one to consider before you proceed with the project. The intended function of your sun space will, in large part, determine the location, size, optional features, and total cost of the project. The more time you spend planning in advance—before you finalize your plans or begin building—the fewer unpleasant surprises you are likely to experience later on.

The functions of sun spaces fit into three main categories: greenhouses for growing plants, living areas, and solar collectors for heating the home. Of course, many people combine two or three uses into their sun space. By incorporating additional features, a sun space intended for one function can serve double—or even triple—duty; but, remember that it will be most efficient when used exactly as it's designed. That's why it's so important to carefully consider exactly what you want, especially if solar collection tubes need to be installed in the walls. Then, if you want to make a modification later, such as a shading system, the original design should be able to accommodate that intention. Some additional features you may want to incorporate at a later date include: for a

Plan your sun space carefully so it serves you and your family well. This dining room makes every meal-time very special.

© Robert Perron

solar collector, solar panels, water tubes, and solar storage material; for a greenhouse, plant shades; for a living space, a shading system. Note that a shading system, important for a greenhouse and a living area, is easily installed at a later date.

The Sun Space As Greenhouse

Nearly every sun space functions to some degree as a greenhouse. The light, heat, and airiness of a sun space combine to create a natural place for you to try out your horticultural skills. A few hanging plants or a dramatic palm tree or two are lovely natural touches that fit almost every sun space. But some people have the desire to do more than just dabble with indoor gardening—they have designed and built their sun spaces to serve as true greenhouses where they can raise and grow any number of exotic species. Owners of sun spaces used primarily as greenhouses enthusiastically describe the relaxing benefits of working with plants. They enjoy observing the greenery's natural cycles, they gain pleasure from raising seedlings to mature plants, and they look forward to the experience of immersing themselves in nurturing and caring for other living things. Truly ambitious individuals even grow food year-round in their sun spaces. Healthy plants, in turn, im-

prove the atmosphere in the home, not only by their beauty, but by replenishing oxygen in the air. However, even the most committed greenhouse enthusiasts admit that there can be problems associated with their hobby.

Sun spaces designed for the optimum growth of plants, which require fairly precise and relatively constant levels of humidity, light, and temperature, are not particularly comfortable for people. So, if you decide to create a sun space primarily as a solar collector,

don't expect also to use it as a living space or a greenhouse.

Let us now consider the two most important requirements for a successful sun space designed as a greenhouse—proper levels of light and temperature.

Light

Plants require high, consistent levels of light for photosynthesis—the essential life process by which they convert light into food—to take place.

Courtesy Appropriate Technology

Above: Even those sun spaces used primarily as living areas can accommodate some plants.

© Vern Green

Above: A bright, cheery room with a combination of overhead and vertical glazing; this sun space really seems to bring the outside in.

During the summer, when the sun is directly overhead, long days provide enough light for most plants to grow. But during the winter, when the sun is at a lower angle in the sky and daylight hours are considerably shorter, plants need all the light they can get. That's why greenhouses should be built facing south, or as close to it as possible, to maximize the amount of winter sunlight. Such greenhouses should also be constructed primarily of glass in order to take advantage of high levels of diffused light. Overhead glazing, too, is an essential element, especially in areas where cloudy conditions prevail, as it maximizes the amount of light available. There are several other ways to maximize the light in a sun space designed for use as a greenhouse; they will be discussed in greater detail in Chapter 5.

Temperature

Optimum plant growth occurs in temperatures ranging from 50 degrees F to 85 degrees F (10 to 29 degrees C). Some plants are more tolerant of temperature variations than others, and some can occasionally weather an extreme, but outside of this basic 35-degree (19-degree) range, the long-term survival of any given plant is doubtful. Therefore, it's essential that a greenhouse be built with these temperature requirements in mind. Consider the following examples: If a sun space will potentially overheat in the summer, you'll need to either invest in a shading, light-reflective glazing, or to reconsider the location of the greenhouse. Similarly, if a southern-

Courtesy Appropriate Technology

© Jerry Howard/Positive Images

Above: Whose day wouldn't get off to a great start in such attractive surroundings. With a lovely breakfast room such as this one, you may never miss breakfast again!

Left: The airy look of garden furniture complements the greenery in this sun space. A shading system helps regulate both temperature and light.

facing greenhouse constructed primarily of glass is subject to bitterly cold westerly winds, you'll need to modify your plans, perhaps by heating the greenhouse (thereby negating any solar energy benefits) to keep the temperature high enough to allow plant growth.

Other less obvious considerations of the sun space as greenhouse include the need for good air circulation to prevent too-humid conditions, the problem of insect infestations, and the fact that greenhouse gardening, as relaxing as it may be, is also a time-consuming hobby. You hardly want to construct a room that creates more demands on your busy schedule than you can handle. A sun space should be a place you can fully enjoy.

The Sun Space As A Living Area

What home owner can't use a little more room? Adding a sun space can turn unused or underused spaces into irresistible living areas. Consider the space in your home: Is there a room that could use brightening? Imagine the spare bedroom completely redone as a sunny hobby room, full of white wicker furniture, thriving plants, and enough open space to allow you to read, listen to music, exercise, or pursue any number of favorite activities. Perhaps you would rather turn the family room into an elegant entertainment area, complete with hot tub, wet bar, and stained-glass panels. Let your imagination soar as you consider the possibilities.

As you imagine all your options, you'll eventually have to take into account certain practical factors that will make your sun space a comfortable living area. Light and temperature, furnishings, and special features should all be carefully considered.

Light and Temperature

The light level in a sun space designed as a living area is not nearly as important as it is for a sun space designed as a greenhouse or solar collector. Therefore, facing south is not a requirement; instead, you might consider a northern,

eastern, or western exposure for your living area. What is important, however, is maintaining a fairly constant temperature in the narrow range of the human comfort level. Although we tolerate wide temperature extremes in the outdoor environment, most people attempt to keep their indoor temperatures near a constant 75 degrees F (24 degrees C), whatever the weather. But maintaining a near-even temperature in a sun space cannot be done without certain modifications. Awnings, reflective glazing, shutters, or shades are virtual requirements for dealing with direct early-morning, midday, or late-afternoon glare, all

common problems that occur throughout the year. Maintaining a constant temperature is not just a seasonal problem but a daily one. The daytime warmth of a sun space may become uncomfortably cool in the evening, even in the summertime. Similarly, the sun space that is delightful in the warm spring and summer may become a dreadfully cold place in the cold of autumn and winter, especially if it faces a direction other than south. Modifications may help, but in bitterly cold climates, the most prudent decision you can make might simply be to close off the sun space during the coldest months.

Courtesy Appropriate Technology

© Peter Paige/design by Rona Levin

Furnishings

The furnishings selected for a living area should be evaluated on the basis of their ability to withstand high levels of light and humidity. Certainly, furniture should be comfortable; although you want it to be sturdy in order to stand up to the sun and heat, it should also convey an airy feeling. Garden furniture is a good choice, as is wicker, rattan, and wrought iron. Bright throw pillows and foam pads soften the look of the room, as do carefully selected plants.

The flooring you choose for your sun space is very important. You'll want to buy a low

Opposite page: An elegant approach. This quilted shading system allows the option of privacy or a lovely view. Shading systems also help to control the temperature in your sun space year round.

Above: Sturdy fabrics and bold patterns go well in a casual sun space. Overstuffed cushions and comfortable chairs soften the look of this room and make it an inviting place to congregate and relax.

Courtesy CSI Solaroom

maintenance, attractive floor that is resistant to both heat and light. Wall-to-wall carpeting, of course, is not a desirable choice. Think instead about natural materials—slate, ceramic tile, brick, marble— which are not only attractive and easily maintained but also efficiently store heat. Small areas can be covered with throw rugs or straw mats without significantly affecting the heat-absorbing qualities of the floor.

Special Features

Consider the special features that you would most like to include in your sun space to create an inviting, beautiful room. You might want to install a hot tub, spa, or pool, a fountain, an eating area, a

wet bar, or even a luxurious bathroom. If these extras capture your imagination, now is the time to dream. You may decide, for example, that you want to install a hot tub or spa. That decision will affect every other decision you make. You'll need to make sure that the foundation is strong enough, and that you have the proper plumbing to accommodate the unit. You will also need a fan to ventilate the room and a drain in the floor. To maintain the temperature level and to prevent excess evaporation from the hot tub or spa, you'll need to protect the unit with a tight-fitting, removable cover. The type of flooring, other furnishings, and glazing materials will be affected by your decision to include a hot tub or spa in your sun space.

The Sun Space As Solar Collector

During the energy crisis of the late 1970s, solar energy was touted as the wave of the future—a non-polluting, ever-renewable, cheap, and efficient source of energy. A decade later, the idea of using energy from the sun no longer seems like a far-out idea. Indeed, many suburban homes throughout the country now have solar-collecting panels attached to their roofs. And in many areas of the country, sun spaces have been constructed with the primary purpose of providing some or all of the heat for the home.

How does a sun space serve as a solar collector? Sim-

Opposite page: Windows and walls, all constructed of glass, bring the beauty of nature close to home.

Left: A simple, angled structure, especially a south-facing one, can be an efficient, effective energy-gatherer, resulting in lower fuel bills.

ply put, the unit collects the light from the sun, which it then stores and distributes as heat. A number of natural principles are involved in this process of heat storage and movement.

Heat Storage—Solar collectors must be designed with materials that will absorb the heat of the sun, then efficiently transport it throughout the house. Masonry floors and walls absorb and retain heat, as do specially designed water columns, high-tech phase-change salts, and chemicals which are embedded into interior walls during construction of the sun space.

Heat Movement—Once the heat is stored by the unit, it must be distributed to other areas of the house. The most efficient methods take advantage of the natural process of heat movement, primarily convection and conduction, both of which occur at the molecular level.

Convection is the transfer or movement of heat through fluids (including the air). **Conduction** is the transfer or movement of heat through a solid, or from one solid to another.

A solar collecting sun space takes advantage of these natural processes to distribute stored heat to nearby rooms. In the air flow created by convection, warm air rises, while cooler air moves in to take its place, and is thereby warmed as well. The use of fans or vents can assist this natural air flow. A solid wall made of brick, slate, or other masonry materials not only absorbs and retains heat, but through the process of conduction helps distribute it as well. A solar collecting sun space designed with a masonry back wall or the addition of water storage tubes separating it from a family room, for example, will transfer heat collected in the sun space to the family room through conduction through the wall.

There are several important factors to take into account before you plunge into building a sun space designed to be used as a solar collector. Consider the following: existing energy efficiency, location and orientation of the sun space, and optional extras such as reflector panels that will enhance the performance of the solar collector.

Energy Efficiency

Obviously, before you consider adding a sun space to be used as a solar collector, you must first determine the existing energy efficiency of your home. It makes no sense to add a sun space for energy savings if you haven't already made your home as well insulated and weather-tight as you can.

Location and Orientation

The location of the sun space is critical when considering building one primarily for solar collection. The glazing should face south—no more than 20 degrees to the southwest or northeast—for the room to collect the maximum amount of solar radiation. Its position is especially important in the short daylight hours of winter.

Equally important is the orientation of the sun space to both the inside and the outside of the house. Where will the heat gathered by the sun space be used? The answer to that question will greatly influence its placement in relation to the other rooms of the house. Will the sun space be used to heat a common living area, such as a family room, a kitchen, or a recreation room? For maximum energy efficiency you will want to take advantage of the natural movement of heat, so you'll want to build the sun space near the rooms to be heated.

Courtesy CSI Solaroom

Above: Carefully consider the optimum placement of your sun space. It should be scaled to the existing structure of your home and properly oriented—both inside and out.

Right: The solar collector, a simple solution to meeting energy needs, is more popular than ever.

Courtesy Pacific Coast Greenhouse

After considering the inside placement of the sun space, consider also its orientation to the outside of the house and to the sun. If the sun space is tucked in on the south wall, with one or two endwalls protecting it from winter winds, it will be considerably more energy-efficient than one tacked on to a south wall without similar protection.

In addition to the benefits of a solar collecting sun space in the storage and transfer of heat, it has the ability to prevent heat loss from the outside wall of the house. How much money will you save if you install a solar collector? It's difficult to predict, but you can maximize the unit's efficiency by carefully considering its practicality in terms of the climate and weather patterns in your area, how many modifications you're willing to make, and how committed you are to the entire notion of solar energy. Obviously, you'll have better results in a sunny location like the Southwest than you will in the cold, cloudy Northeast.

Above: **A simple, south-facing sun space with overhead glazing is the most efficient solar collector. It takes full advantage of the changing angle of the sun's rays.**

Right: **An example of a well-planned sun space that beautifully complements the rest of the home, both in terms of materials and design.**

Planning The Site For Your Sun Space

Once you've decided what type of sun space you want to build—greenhouse, living area, or solar collector—you'll need to determine where it should be built. As we've discussed, there are several important considerations that factor into determining the optimum location of a sun space.

Step One: The first step is actually to draw a site plan of the house and lot. It's far simpler to get it all down on paper than to try to visualize your home and surrounding lot. The site plan should be drawn to scale on graph paper to ensure the most accurate presentation of the site. Be sure to include the living areas of your home, nearby trees and buildings and the shadows they cast, the orientation of the house in relation to true south, the direction of prevailing winds, and the angle of the midday sun in both summer and winter. (For more information on how to draw a site plan, see page 29.)

Step Two: Determine the size and location of your sun space. On your site plan, sketch the new addition in various locations, keeping in mind its primary function. Work carefully on the design, orientation, and placement of your sun space while you carefully consider the direction of the wind, the climatic conditions in your area, and how you can maximize the energy efficiency of the unit.

Step Three: Consider how the addition of a sun space will affect the rest of your home. How will it change traffic patterns and views within your home? Will the unit fit your home, both in style and scale? Is it compatible with the roof pitch? Most sun space owners wish they had decided on a larger unit; that's a choice that should be considered before you begin building.

Step Four: Study local zoning requirements and investigate what kind of approvals (city or town regulations or your condominium board, for example) will be needed before you begin construction. How close to your property line can you build? How large can the building addition be? Plan now for electrical, plumbing, and ventilation requirements.

The key at this stage is in planning carefully for every foreseeable circumstance. A little extra time and thought now will prevent regrets and problems later on.

The changing angle of the sun's rays must be understood and carefully applied when you consider adding a sun space. As you may remember from your geography courses, the earth spins on its axis at a 23½-degree angle as it re-volves around the sun. It is that tilt that is responsible for the changing of the seasons; were it not for the tilt, half of the earth would experience perpetual winter, while on the other half it would always be summer. During the earth's year-long elliptical path around the sun, the angle of the sun's rays varies by 47 degrees. In the summer, the noonday sun is nearly overhead, which contrasts considerably with the low angle of the noonday sun during the winter. Your sun space should be designed and built with these natural processes in mind to maximize solar radiation in the short days of winter and avoid overheating in the summer.

A STEP-BY-STEP GUIDE
How to Draw a Site Plan

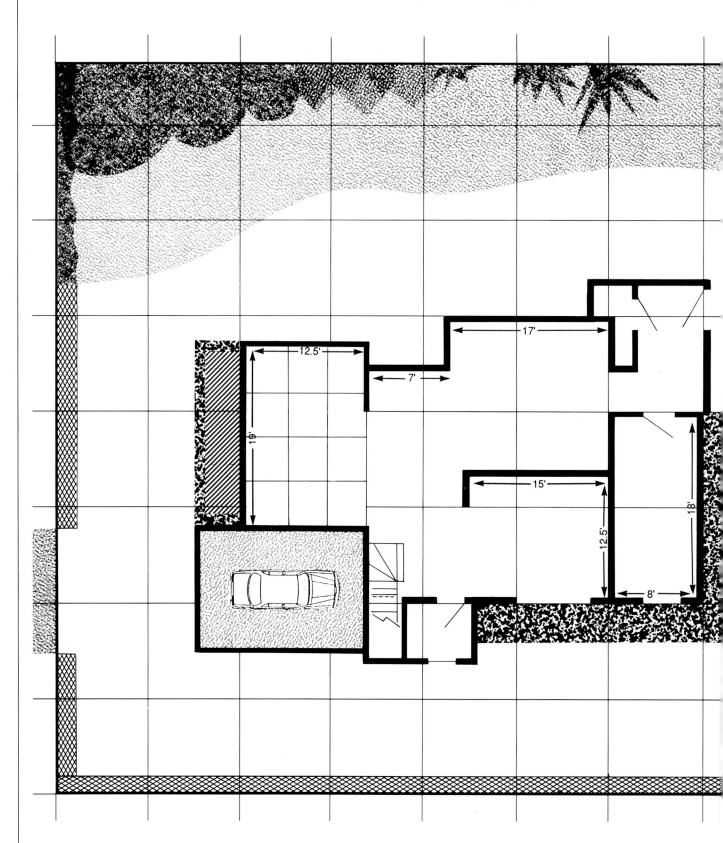

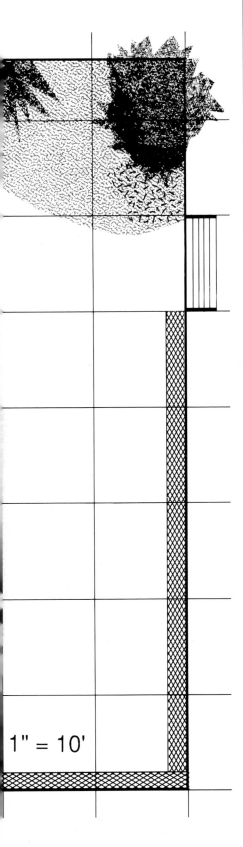

1" = 10'

The first step in the nuts-and-bolts planning of your sun space is to do more than visualize it—you must commit your ideas to paper. To do so, you need to create a site plan to accurately determine where to place the sun space in relation to the physical factors of your home.

1 Gather the materials you need—graph paper, sharp pencils, a rule, a good measuring tape, and a couple of willing helpers—and you're ready to get to work.

2 Roughly sketch out the floor plan of your home, marking doorways, halls, closets, and windows. Don't let yourself be intimidated; just visualize a bird's-eye view of your home and plot it out on paper.

3 Measure the dimensions of each room in your home, and carefully sketch out the floor plan on the graph paper. Keep the dimensions in correct proportion in the sketch—let one inch equal ten feet, or any other scale that works right for you.

4 When you're finished plotting the entire floor plan as accurately as possible, sketch in the additional features that are part of your home site: deciduous and evergreen trees, the lot line, the location of the street, the garage, any patio, deck, or other existing or planned structure that might cast shadows or otherwise interfere with the placement of the sun space. Also note compass points, prevailing winds, and any other details that seem to be relevant to you.

Make several copies of the site plan so that you can sketch the sun space in different locations; visualize your options and, after trying several on paper, make a realistic decision about how and where your sun space will fit into your home.

CHAPTER 2

Basics About Materials

Now that we've discussed the many functions of sun spaces, it's time to examine their structure in greater detail. In order to make a wise decision in the selection of materials, it's important to have a basic level of understanding of the many options available today. You have a wide range of choices, depending on your needs, in the type of glass, framing materials, foundations, and other features to be built into the structure of your sun space.

Glazing Materials

To the trade, the term "glazing" includes the various types of glass, acrylics, and plastics—transparent or translucent materials—used to transmit light in places such as windows, doors, and sun spaces. Glass is superior to various plastics in many ways: It transmits light better, it is less prone to scratching and discoloration, it insulates better, it is more transparent, and it distorts views less than plastic.

When considering glass, it's important to consider two key factors: the structural ability of the glass or glazing to withstand stresses and loads, and the extent to which the material transmits light and heat. When glazing materials are carefully chosen with the above factors in mind, the enjoyment of the sun space can be increased immeasurably.

Structure

The type of glass that is found in the windows of the typical home is known as **annealed** glass. Because of its tendency to shatter into sharp fragments on impact, it cannot be used in doors or overhead panels. Annealed glass is often found, however, in the vertical panels of greenhouses built in mild climates.

Heat-strengthened glass, like annealed glass, shatters into sharp fragments on impact. But because of the heating, reheating, and cooling process used to construct this type of glass, it is twice as strong as annealed glass, making it more resistant to shattering on impact or when stressed by heat loads. It can be used in overhead panels, but only when laminated or screened for protection.

Four times stronger than heat-strengthened glass, **tempered** glass is highly resistant to impact and heat loads. When this glass breaks, it forms small, rounded fragments, not the dangerous shards of annealed or heat-strengthened glass. Tempered glass is considered a safety glass for use in doors and panels, but it cannot be used overhead without the added protection of a wire screen because of its tendency to fall out of its frame if it breaks.

A beautiful combination of vertical and overhead glazing provides a window on the world from this lovely room to the great outdoors.

© Robert Perron

Above: **Overhead glazing and curved framing create a bright room with an unobstructed view. Appropriate glazing should be carefully selected for safety and comfort.**

Right: **The right glazing can make the difference between a successful and an uncomfortable sun space. Double-glazing is a standard choice, and tints or coatings can help control heat gain, which is especially important in sunny areas.**

To solve the problems of glass shattering or breaking, especially in overhead uses, **laminated** glass has been developed. As the name suggests, laminated glass consists of two pieces of glass permanently bonded together with an inner layer of plastic film between them. The double thickness of glass, which can be annealed, heat-strength-

ened, or tempered, plus the plastic interlayer causes the glass to be even more impact- and heat-resistant than just a single layer of glass. If the glass should break, the plastic inter-layer holds it in place and prevents it from evacuating its frame; this is especially important in overhead uses.

Although building codes vary, most require that laminated glass be used when placed overhead, others require tempered glass in places such as sliding doors, where the possibility of breakage is increased. Check with your builder and the local building codes to determine the best type of glass for your needs.

Light and Heat

The next consideration in the glass structure is the amount of light and heat transmitted by the glazing system. This area involves not only the number of layers of glazing

appropriate for your sun space, but the special coatings that can dramatically affect the heat and light collection of the glass.

Single-glazing, that is, a single layer of glass, is usually found either in traditional, unattached greenhouses or in sun spaces designed for maximum energy transmission in very sunny climates with little or no need for insulation.

The industry standard for energy-efficient, comfortable sun spaces is a system utilizing **double-glazing**—that is, two layers of glass with a layer of

air trapped between them. Such a system takes advantage of the natural insulating properties of air space, resulting in a 50 percent reduction in heat loss to the outside. Because of this, double-glazing is especially desirable in cooler climates; in some areas of extreme cold, triple-glazing systems are used, although this does become very expensive. To reduce the problem of condensation between the layers of glass, the panes are set into aluminum or wood frames with a special chemical drying agent.

Thanks in large part to space-age technology, exotic, specific-purpose glass coatings have been developed to create glass surfaces that are more efficient solar collectors or reflectors. Such qualities are extremely important both in hot climates, where dealing with excessive heat gain is a real problem, and in cold climates, where similar problems occur in dealing with heat loss. The right choice of glass coatings can make the difference between an intolerably hot or cold room and one that is comfortable all year round.

Low emissivity (low e) glass. This metallic coating, developed in the late 1970s, is particularly effective in cooler climates. It is designed to reflect radiant heat back to its source: out of the sun space toward the sun in the summertime, and back into the sun space in the wintertime. The coating transmits visible light, but blocks ultraviolet rays, resulting in less glare and

Courtesy American Olean Tile Co.

considerably less fading of materials inside the sun space. Low e glass is warmer to the touch than uncoated glass, which reduces condensation —an especially important consideration when the sun space encloses lots of plants, a pool, a hot tub, or a spa.

Tints. There are a number of tints that can be applied to the outside of the glass to shade or reflect heat and light. Among the most common shades are bronze, green, blue, gray, and black. One drawback to tinted glass is that it reduces the amount of light transmitted to the sun space by about 30 percent, while heat gain is even more significantly reduced, up to nearly 90 percent. However,

tinted glass can be quite desirable, especially in very warm, sunny climates where excessive heat gain can be a major problem.

Also falling into the category of tinted glass is **reflective** glass, utilizing tints of silver or gold, which is also quite effective in places with lots of sunlight and heat. Whereas tinted glass blocks light, reflective glass reflects the light back toward the light source.

As we've seen, there are many choices to be made in the selection of glazing materials for your sun space. Working with an experienced builder or sun space dealer will be extremely helpful in making decisions that are best suited to your needs.

Right: **Aluminum has been used to frame the glass in this sun space. Because aluminum is both lightweight and maintenance free, it is the perfect material to use in sun spaces.**

Below: **Today, sun spaces can be constructed to be compatible with any style home.**

Courtesy Four Seasons Greenhouse/System 4

Courtesy American Olean Tile Co.

Framing Materials

The two most common framing materials used in constructing sun spaces are aluminum and wood. Each system has both advantages and disadvantages.

Aluminum

Lightweight, modern, and virtually maintenance-free, aluminum frames allow for sophisticated detailing and a light, slim look in a sun space. Aluminum units are very popular; they are also somewhat less expensive than sun spaces constructed from wood. Because it is relatively impervious to high humidity, an aluminum unit is a good choice for a sun space used primarily as a greenhouse; if used along with a hot tub, pool, or spa, the aluminum surfaces should be specially treated to resist chemical damage. Anodized aluminum frames are available in a range of colors, but black, white, and dark brown are the most popular. The aluminum can be custom-painted to give your sun space a more integrated look.

To prevent the formation of condensation, which can cause leaking and dripping in the aluminum-framed sun space, the bars, which form the basic structure of the aluminum frames, are designed with internal drainage systems that drain the condensation outside. Thermal breaks—systems of gaskets that stop

the conduction of heat or cold between the inside and outside metal, thus preventing the potential problem of extremely cold framework during the winter—and concealed fasteners reduce heat loss in aluminum units.

Wood

Sun spaces constructed with wooden framing materials have a warmer, more traditional feel. They are especially attractive in older homes, or those where it's desirable to create a classic design. The wood used in the construction of a sun space requires maintenance to resist moisture and minimize the inevitable process of weathering. Still, the attention required to maintain the condition of the wood—sealing, staining, periodic refinishing—is fairly minimal, and may be offset by its warm look. Some consider the more weighty appearance of wood a disadvantage, but others like the natural look, particularly when combined with a lovely view of the outdoors.

Unless the sun space is to be used near a source of water—a significant number of plants, a pool, hot tub, or spa—or there are specific design requirements, the decision about the use of aluminum or wood framing materials is largely a matter of personal esthetics. Some manufacturers have recently combined the two materials, creating a sort of hybrid unit that features the best of both worlds—the lightness, durability, and low maintenance of aluminum hardware combined with the beauty of natural wood.

Courtesy Regal Manufacturing Co.

Left: An aluminum and glass sun space is a simple, attractive home addition that is far more dramatic than a similarly sized traditional patio or enclosed porch.

Above: Wooden framing and beams provide an outdoorsy, warm feeling that is especially attractive when combined with overhead glazing.

© Robert Perron

Above: **A well-constructed sun space starts with a solid, insulated foundation.**

Right: **This handsome unit is well-integrated into the overall plan of this large, traditional home—the result of a careful assessment of real and desired factors.**

Foundations

Starting from the ground up, a good, stable foundation that both supports and anchors the structure is essential for your sun space. It's imperative that the foundation be solidly constructed, because a sun space, with its high proportion of glass, cannot shift or settle without (literally) shattering consequences. Hilly sites, areas with poor soil drainage, the earthquake-prone West Coast, and other factors, including heavyweight options such as hot tubs, require special consideration in the foundation design. Because it is such an important component of the sun space, an experienced, reputable builder should definitely be consulted.

All foundations should be insulated. Those that come in direct contact with the ground should be insulated with moisture-resistant materials at the frost line. This is especially important if the sun space is to be used for solar collection, because a well-insulated, solid concrete foundation functions as a thermal mass.

The simplest, most common type of foundation is one constructed from **continuous masonry**—usually poured concrete, concrete blocks, bricks, or stones—which run under the entire perimeter of the structure. Such a foundation is strong, solid, and easily insulated, although it does require a considerable amount of excavation, particularly in extremely cold climates where the frost line extends deep into the ground. Because of the preparatory work, this foundation tends to be expensive.

Pier foundations, although not quite as substantial as those constructed of continuous masonry, are still strong enough to adequately support a sun space, and are best for hilly sites or where the sun space is otherwise constructed above ground. Pier foundations are less labor-intensive and require fewer materials than those that need excavation; therefore, they can be built for less cost than masonry foundations.

Wood foundations constructed of specially treated materials to resist moisture and insects, are simple, easy to construct, and have become very popular in recent years. With this type of foundation, posts are sunk deep into the ground to provide sufficient support; the posts can also be used to provide vertical support for sun space units with kneewalls.

Additional Considerations

As we've discovered, there are numerous options available in nearly every phase of planning and building a sun space—walls, insulation, the ventilation system, the slope and configuration of the glazing—all must be determined before the actual building begins on your space.

Walls

By definition, a sun space consists primarily of glass and framing materials, but kneewalls, also called basewalls, and endwalls are very common—and often preferred to extending the glazing to the foundation. Both the smaller kneewalls (short walls built

Courtesy Four Seasons Greenhouse

under the bottom of the glass in a sun space) and the more substantial endwalls (walls on the ends of the sun space) offer a certain amount of privacy and additional insulation; they may provide an extra measure of esthetic appeal as well. Sun spaces constructed in very cold climates are more energy-efficient when built with insulated kneewalls, basewalls, and endwalls; endwalls are especially good when used as buffers against blasts of bitter winter winds. Endwalls can

mask an unattractive view; kneewalls and basewalls can protect children from colliding with near-the-ground glazing when they play outside; and all three types provide endless opportunities to enhance the esthetic appeal of a sun space. Consider, for example, the integral appeal of a kneewall or basewall built of brick to match the exterior of a traditional home. Imagine the beauty of incorporating a window seat built at kneewall level into a sun space, both

inside and outside of the structure. Using a solid or partly solid wall in the sun space can also simplify the installation of electrical outlets, an important consideration, and one you will appreciate later.

Ventilation

Adequate ventilation in your sun space can make all the difference between total enjoyment and complete disaster. Proper ventilation in the summer will allow you to use the room without feeling overheated, especially if the unit is designed to take advantage of naturally occurring afternoon breezes. Screened windows and doors, vents, and fans are all simple systems which, when carefully placed, provide a considerable level of comfort in your sun space and allow you to fully utilize the space throughout the year.

Overhead Glazing

One of the most appealing attractions of sun spaces is the presence of overhead glazing, allowing you to gaze up into the heavens while enjoying the protection and comfort of the indoors. But sun spaces that incorporate overhead glazing can easily overheat, especially during the summer, in western- and eastern-facing units that absorb early-morning and late-afternoon sunlight. Because of the low angle of the sun's rays during the winter, overhead glazing is not an advantage in solar collection; in fact, overhead glazing does little to help retain heat in the winter. With these considerations in mind, there are many ways to deal with the question of the right amount of overhead glazing to meet your needs. Overhead glazing may be built into the entire unit, it can be incorporated as just one section of the overhead structure, or it can be eliminated altogether in a sun space built with vertical glazing only.

Left: **Imagine star-gazing into the heavens from the comfort of the indoors. Overhead glazing can be used quite successfully and dramatically, particularly when light is filtered by outdoor foliage.**

Below: **In this treatment, overhead glazing is combined with attractive overhead fans for comfort and visual appeal. The result: a cool, enjoyable room.**

Courtesy Paeco Inc./Sol*Area

Shading Systems

Virtually every sun space will need some sort of shading to protect it from overheating in the summer. Heat gain is often a problem in sun spaces, but it can be dealt with in a number of ingenious ways. Perhaps the simplest, most intriguing solution is to place a substantial amount of plant life outside the sun space, primarily in the form of deciduous trees. Because the trees lose their leaves in the winter, they allow the sun to penetrate the sun space unobstructed; dur-

ing the summer, their abundant leaves not only shade the sun space, but provide a lovely view as well. However, if the growth of such trees becomes too dense, they will shade the sun space too much; a careful balance must be determined.

More expensive than trees and other forms of vegetation, but still a simple solution, is the use of adjustable outdoor blinds, such as bamboo slats, or vinyl or matchstick blinds. These blinds can be raised and lowered as needed, and they can cast pretty shadows as the

Right: **Contemporary and classy— the sleek design of an ultra-modern sun space.**

Below: **The soaring grace of curved glazing adds immeasurably to this attractive sun space used primarily as a luxurious living area—complete with indoor spa.**

Courtesy Comfortex Corp./Duette Smart Shade

Courtesy Lindal Cedar Homes, Inc.

softened light enters the interior of the sun space. Outdoor blinds sustain quite a bit of wear and tear, and may have to be replaced each year, but they tend to be relatively inexpensive and their functionality makes them worth the price of replacement. Shades constructed of more durable materials, such as cedar or aluminum, are more expensive, but will last longer.

For those who find blinds inconvenient, unattractive, or impossible to fit to their unit, another outdoor option to consider is the use of attractive awnings. The bright, pretty striped awnings, similar to those found in outdoor cafes and sidewalk displays, can be very attractive and functional when used to protect a sun space. They are a more expensive alternative, but they tend to last longer than roll-up blinds. Indoor shading systems range from the very simple—roll-up blinds attached inside—to the complex—an automatic system using quilted shades attached to rollers. We'll discuss the many luxurious options in shading systems that you can choose from in greater depth in Chapter 6.

Eaves

The place where the vertical glazing meets the roof line in a sun space is known as the eave. There are basically two major choices in the shape of the eave for your sun space: curved and straight. Curved eaves seem to soar; they dramatically loft skyward, offering an unobstructed view of nature's beauty. They also offer the additional benefit of their aerodynamic shape, important in areas prone to high winds. Straight eaves, on the other hand, provide a sleek, contemporary look that fits in well with the design of many of today's modern homes. Both choices are adaptable to a number of styles, roof lines, and, importantly, the roof pitch of a home. Kits featuring both curved and straight eaves are widely available. It's important to consider the look of the unit, from both the interior and the exterior, when you are deciding on the configuration of the eaves you want to incorporate into your sun space.

Kits

There are many kits now on the market that can best supply what you're looking for. Although some sun space manufacturers have been in business for more than a century—most notably Lord &

Courtesy Four Seasons Greenhouse/System 6

Above: **A well-built, comfortable breakfast room, and so much more. It's a place where the entire family will want to spend many hours at a time.**

Burnham, which was founded in 1856—most of the boom in the business has taken place since the energy-crisis days of the early 1970s.

Attached sun spaces were first used as solar collectors in an attempt to use the energy of the sun as an alternative to the burning of fossil fuels. Tax credits made the building of such sun spaces advantageous, not only in terms of energy savings, but financially as well. Unfortunately, a few manufacturers jumped into the business with shoddy products, made a quick buck, and got right out. That disreputable behavior left a number of owners with units that were far less acceptable. Many experienced problems with leakage, breakage, and generally poor workmanship.

But, with this second generation of sun spaces on the market, improved materials and engineering, and the virtual elimination of tax credits for solar energy, the early problems with here-today-gone-tomorrow, make-a-quick-buck companies and their poor-quality construction have largely been eliminated. Today's buyer can research the market and make wise decisions about the space that will be best for his or her home.

As materials improved, individuals began to rethink the function of sun spaces. No longer primarily designed as solar collectors, today's sun spaces are frequently constructed to be used as living space, although greenhouses and solar collectors are also very popular. Fortunately, manufacturers of sun space kits have designed lovely units that fit virtually every function you may have in mind for your sun space.

How do potential buyers determine which sun space will best fit their needs? Is a prefabricated kit the best option, or will a custom unit be better? Do you need to hire an architect for an original design, or is it better to try to do it yourself?

Naturally, no one can answer these questions for you, but here are some simple guidelines to follow to help you make your decision.

As previously mentioned, the best way to determine exactly which sun space will work best for you is to decide the function and structure that you most desire. Probably the best way to become familiar with the various kits on the market is to obtain and read the literature of the sun space

Below: **Furnishings of wicker, cane, and willow are a great accent in this natural-feeling sun space.**

Courtesy Lindal Cedar Homes, Inc.

manufacturers. With more than fifty different sun space companies, each offering several options, this early research stage can be a daunting task. But it's also well worth your time to examine the colorful brochures and read the information about the various companies and their wares. As you pore over the catalogues, you'll become even more enthusiastic about building your sun space and it will be much easier to decide what you want.

Another good way to make decisions during this early stage is simply to visit the dealers in your area that have sun spaces on display. While there, you can examine the structures first-hand and get a very good sense of the quality, durability, and esthetic appeal of the model that most interests you. Additionally, you'll have the opportunity to talk with the builder or dealer, who, if knowledgeable about the product line, can provide valuable insights and expert advice about the best sun space for you— and what options you might desire, particularly in the area of glass selection for your geographical location.

Another option is rather adventurous—you might want to explore a bit. Drive around in areas where people have already had sun spaces built; notice the configuration of the unit, the type of glazing, whether it's tinted or not, how it blends in with the rest of the home. You may even walk up to the door, introduce

yourself to the owners, and ask a few questions about how they like living with a sun space. Chances are they'll be pleased to offer you a tour, and will be happy to share their experience with you. There's no better reference than a satisfied customer; the opposite is also true.

If you keep your eyes open, you'll notice that many restaurants (including several fast-food franchises), banks, shops, and other commercial enterprises have incorporated sun spaces into their buildings. Don't hesitate to visit those places and carefully examine

the structure while you're there. You'll learn valuable information about the interior of the sun space; take note of the use of shades, plants, ventilation, and any other modifications made to the unit. During such visits, you may also gain ideas that you'll want to incorporate into your own sun space. For example, a bustling Southern California restaurant has extended its floor space with a large, curved-eave sun space, and has replaced several of the stock sections of vertical glazing with stained-glass windows that sparkle beautifully, night and day. An-

Left: One excellent way to research sun spaces is to view the ones in your neighborhood. Notice their design, special features, and the little details to help you plan your own.

Below: Examine existing sun spaces for their compatibility with the structure of the home, their warmth, and their furnishings. Perhaps you can even arrange a tour with the owners.

Courtesy Regal Manufacturing Co.

Courtesy Four Seasons Greenhouse/System 4

other popular restaurant, this one in the rainy Northwest, has installed a sun space that incorporates a substantial amount of overhead glazing. Diners particularly enjoy the opportunity to eat in an almost-alfresco environment, despite the frequently inclement weather.

Don't be afraid to ask questions about these sun spaces as you examine them with an eye toward building your own. Many of the commercial applications of sun spaces are simply standard kits, available on the retail market. After you've studied your brochures, you'll not only be able to identify the manufacturers of the most popular units, but you'll also be well-informed about the actual comfort and use of the particular sun spaces you might be considering for your own home.

As you investigate the kits on the market, you'll probably find several that would be just right for your taste, size, and space available, as well as your wallet. The question of whether you should construct your sun space from a kit or have one custom-built is an important one. It may, however, be a moot point; up to 95 percent of all kits need to be customized in some way, so that they will precisely fit the site for which they've been selected. In most cases, an architect or builder who is familiar with the design and construction of sun spaces should be consulted at the earliest stages. Their expert advice will prove invaluable in every aspect of planning, from determining the placement of the unit to finishing its interior.

The great benefit of prefabricated kits is that they are precision-engineered at the factory; they are fashioned of quality materials and are fairly simple to adapt in terms of size, roof pitch, and desired options, such as windows, vents, and doors. Kits usually take less construction time and tend to cost somewhat less than custom-built sun spaces; good builders familiar with sun spaces can usually modify a kit to fit your home in de-

Before, during, and after the installation of a sun space kit. It may look easy here, but be sure to properly plan for this addition. This unit fits the design of the home and greatly enhances its look.

sign, structure, and esthetic appeal.

The next major question to consider is whether you want to hire a builder for the construction of your sun space, or to do it yourself. A builder recently observed that sun spaces are designed to be showplaces; they're not the place to scrimp and save money. If that's the reason you would decide to do it yourself—to save money— you might want to rethink your decision. Building a sun space is not a casual undertaking. It's at least as complex as any other remodeling project, but because of the exacting requirements of fitting the glass together with framing materials, it can require considerable sophistication and knowledge of the construction process. The typical sun space may consist of up to six thousand separate parts; clearly, constructing a unit is more than just a weekend project. But, if you are a seasoned do-it-yourselfer and are confident of your building knowledge and ability you may opt to build your own sun space. The precut, predrilled, precisely aligned and numbered modular kits have been designed to facilitate their construction. But, if your idea of a nightmare is assembling a child's toy, with part A, part B, *ad infinitum*, then constructing your own sun space is definitely not for you. Instead, you should consult a builder to construct it.

What kind of guarantees/ warranties can you expect when you purchase a sun space kit? As you may suspect, they vary greatly from one company to another, but most provide limited warranties for parts, seals, materials, and workmanship; however, it is important that you know that many kits require dealer installation for the warranty to be effective.

The component parts of your sun space require careful consideration well before construction begins. Remember that you have several options in almost every aspect of sun space construction, and that your decisions will greatly affect your enjoyment of your sun space. But time put into research to help you make informed choices will make the preconstruction process not only informative, but a great deal of fun as well. Be careful, be deliberate, and keep good records, and you'll soon be well on your way to creating the exact sun space that you want and one that you will fully enjoy.

Courtesy Brady & Sun

Courtesy Brady & Sun

A STEP-BY-STEP GUIDE
How to Build a Window Seat

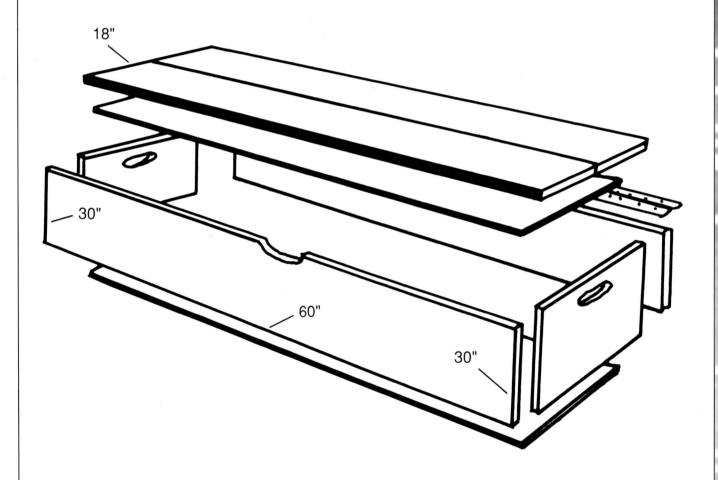

Nowhere is a window seat more appropriate than in a sun space. Basically nothing more than a large box that's sturdy enough to support whatever weight you intend to place on it, a window seat is fairly simple to construct. A few hours of time and fairly inexpensive materials, accurate measurements, and a little labor combine to make a very special addition to your sun space.

Step 1: Decide where to place the window seat within your sun space. For maximum practicality and to allow one to lean back in comfort, the window seat should be constructed between two sturdy walls, preferably on a short endwall of the sun space. Also remember that if you will be constructing a window seat with built-in storage (as we do here) it will need to be ½ inch (1.5 centimeters) away from the wall for it to open.

Step 2: Determine the length, height (kneewall level is good), and depth of the bench. Here, using ¾-inch (1.875-centimeters) plywood we will use the following measurements: length = 60 inches (150 centimeters), height = 30 inches (75 centimeters), depth = 18 inches (45 centimeters). Cut the plywood to the desired dimensions.

Step 3: Beginning 3 inches (7.5 centimeters) from the top edge, drill five pilot holes through the front piece into carefully aligned side pieces, ½ inch (1.25 centimeters) from the side edge at the following intervals: 3 inches (7.5 centi-

meters), 9 inches (22.5 centimeters), 15 inches (37.5 centimeters), 21 inches (52.5 centimeters), and 27 inches (67.5 centimeters).

Step 4: Align the long side of the window seat with the short side of the window seat. Screw securely with 1½- to 2-inch (3.75- to 5-centimeter) screws.

Step 5: Repeat steps 3 and 4 with the opposite side and the backside.

Step 6: For the bottom: Turn the four pieces already secured together, upside down. Drill pilot holes through the bottom of the structure, into the side walls (every 6 to 8 inches [15 to 20 centimeters]), starting 3 inches (7.5 centimeters) in from the corners. Repeat this on all four sides of the bottom. Screw the bottom piece of plywood securely with 1½- to 2-inch screws.

Step 7: Turn the structure right side up. Attach a piano hinge onto the top lip of the back surface so that the spine of the piano hinge is hanging slightly over the back. Make sure that the hinge is opened to an 'L-shape'. Screw the hinge into the backside of the box every 3 to 5 inches (7.5 to 12.5 centimeters). Place top on box, into 'L' shape of the hinge. Screw in from the back, using ¾-inch (1.875-centimeters) screws.

Step 8: Sand the surface and finish the window seat with stain or paint as desired. You may want to cover it with sturdy industrial carpeting for a sleek look or with fabric for a more romantic look. Add a seat pad to run the length (60 inches [150 centimeters]) of the bench, and toss pillows on the window seat for additional comfort.

Step 9: Put the window seat in place (don't forget that the seat must be ½ inch [1.25 centimeters] away from the wall in order to open), sit back with a favorite book, and enjoy your new window seat.

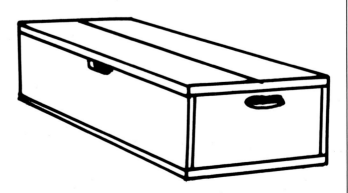

Courtesy Lindal Cedar Homes, Inc.

CHAPTER 3

The Building Process

Courtesy Regal Manufacturing Co.

Above: **Tricky lines and angles are best attempted—and constructed—by experienced professionals. Careful selection of your building contractors is essential.**

The preliminary steps in the building process—finding not only the right unit but the right people to build it for you—are essential in making it as hassle-free as possible. There are all kinds of jokes about how traumatic remodeling your home or building an addition can be; they're not without some truth. Allowing strangers into your home to knock out walls, remove part of the roof, and all-in-all to create an unbelievable mess throughout your home—while charging a great deal of money—is a stressful proposition at best. But taking the time to determine exactly what services you want and who can best provide them—who will create the least amount of disruption in your home and the least amount of anxiety about the quality of their work—is a small investment in time early on that will eventually pay off.

We'll first take a look at the initial research stage, where you study the various contractors and builders, take and judge bids, and finally decide who will work best with you. Then we'll examine the actual building stage and what to expect from it; the payment schedule and time requirements you can expect; and, finally, the option of acting as your own contractor, subcontracting the major jobs and constructing the sun space on your own. Armed with such information, you'll be able to forge ahead confidently.

Initial Research

As previously mentioned, word-of-mouth advertising is one of your best bets in the construction business. Talk with trusted friends and family, and ask present sun space owners about their experiences with their builders and contractors. Find out how satisfied they were with their remodeling or other building projects; ask for recommendations and who or what to avoid. Most people are more than willing to share their experiences, both good and bad. Real-life experiences will tell you much more than any book or company literature.

In addition to talking with individuals, you should check with your local building department, Better Business Bureau, or Consumer Affairs office to research the companies you're considering. Make sure that these companies are licensed by your state, and determine whether they have any complaints on file. What you're really trying to accomplish in this early stage is to determine the quality of the companies' reputations in your community.

Naturally, you'll need to spend some time with the building contractors you're considering. No matter what others recommend, you're the one who has to deal with the

Below: **Time spent researching before building will pay off in the years to come.**

Courtesy Green Mountain Homes

representatives of the company you select. As you become involved in discussions about your sun space, pay attention to telling details, such as how promptly and courteously your phone calls are returned, whether the contractor is on time to appointments, and if you feel you're on the same wavelength. The way you're treated at this stage is a good indication of the company's style, and how you'll be treated later. It is essential that you feel comfortable with the individuals who will be working in your home.

Once you've narrowed your choices down to two or three builders with whom you feel comfortable, you can begin the bidding process. Make certain the contractors know exactly what the specifications are for your sun space, and carefully spell out exactly what's expected of the contractor in the way of cleanup and the removal of debris. At this stage you want to be sure that each contractor bases the estimate on the same quality and amount of materials and labor. For the best basis of comparison, request carefully detailed, written estimates. You want to inform each contractor that you're soliciting comparative bids; that knowledge may influence the estimate in your favor.

Once the bids are submitted, evaluate them carefully. Remember that it does no good simply to pick the cheapest bid—you could run into problems later on. If one bid seems out of line with the others, question the contractor, try to determine the reasons for the discrepancies. A substantially lower bid may indicate low-quality work. The old adage, "You get what you pay for," may come into play.

When you're satisfied with the bids you've obtained decide on the contractor who most appeals to you, based not only on price but also demeanor, style, and attitude. It's a good idea, especially if you're having a difficult time making choices at this stage, to ask for references; talk with a few of the contractor's past clients, and even visit the contractor on a job site. Pay close attention to the response you receive when you suggest this. If a contractor seems less than forthcoming, there may be good reason—and something to hide.

After all this research, you'll be able to make a wise, informed decision that will allow you to be comfortable while the project is under construction. This will ultimately increase the overall enjoyment of your sun space—which is the real object of all of this.

The last step in this initial stage is to draft the actual contract. The contract is the legal document that specifies responsibilities, price, and all agreements made between the parties involved. Your primary concern at this point is to ensure that the contract is very specific and written without ambiguities that could cause problems in case of a dispute later. It may be advisable to hire a construction attorney to review the contract before you sign it.

Your contract should include two provisions: an "assignment clause," which prohibits the contractor from hiring subcontractors without your prior approval; and a "release of lien," which, if subcontractors will be working on the job, eliminates you from liability if the builder, despite all your careful investigation, does not pay.

Left: A simple, straightforward design; this sun space was carefully created to complement the style, size, and materials used in this traditional home.

Below: This graceful sun space required plenty of planning to incorporate its unique features—but look at the results; they're worth every moment.

Building The Sun Space

Courtesy Pacific Coast Greenhouse Mfg. Co.

Above: **A professional job, well-done. This handsome sun space is large, full of light, and very appealing with its simple furnishings and natural surfaces.**

All new building is carefully regulated and must conform to both national and local building codes. The two major areas of regulation are safety and comfort. Safety concerns have to do with the reduction of the risk of injury to individuals, and damage to property, in the event of fire, earthquakes, high winds, or structural failure. Electrical and plumbing systems also fall into the category of safety regulation. Comfort concerns include provisions for such aspects as adequate ventilation, ceiling height, and room size.

Before any on-site work begins, discuss your plans for your sun space with a representative of your local building inspection department. A problem that may occur here is a matter of semantics. Sun spaces may be referred to as greenhouses, solariums, hothouses, solar rooms, or simply room additions. Whatever the term, be sure that the building inspector knows exactly what sort of building project you're planning.

When you apply for a building permit, you'll need to submit plans for your sun space, along with a sketch of your home site. Be prepared to pay for a permit—the fee can be substantial. In some areas, building permits can cost up to ten percent of the total cost of the building project. If there are any zoning problems,

Below: Building codes regulate safety and comfort concerns, ensuring long-lasting enjoyment. Because codes vary from place to place, check with your local official for exact requirements.

based on the determination by the city planner's office or local zoning department, you may have to seek a variance —that is, the official approval to build your project even if it's in technical violation of local regulations.

Before construction can begin, two important steps remain: Obtain the business permit and, for your own protection, discuss insurance provisions with your contractor. Make sure that all of the employees are adequately covered by Workmen's Compensation, or another similar insurance policy, releasing you from any liability if an accident occurs on-site.

When building finally begins, keep in mind a few rules of thumb that may lessen your anxiety and stress during the construction:

© Robert Perron

1 Remember that the contractor is probably working on more than one job at a time, so the crew won't always be on-site when you think it should be. The whole project will probably take more time than you expect; imperfect materials, labor problems, and of course the weather can delay the completion, so be prepared for the inevitable. You may want to plan ahead and schedule the building of your sun space at a time when your contractor is not too busy with other projects.

2 Since time is money, remember that the completed project may run somewhat higher than the original estimate. Stay in close touch with your contractor and discuss any cost overruns as they occur. Don't depend on verbal agreements; put all changes in writing, called "change orders."

3 Contractors vary in their payment requirements, but it's fairly standard to pay for half the cost of the materials when ordered and half on delivery; the labor costs are paid in installments, or when the entire job is completed. It's common to hold approximately ten percent of the total cost as the final payment, delivered when the work has been completely —and satisfactorily—executed. Work out a payment schedule that feels comfortable for you and the contractor.

Courtesy Solar Resources, Inc.

Courtesy Solar Resources, Inc.

Do-It-Yourself

Up to this point, we've assumed that the sun space would be wholly contracted by a builder familiar with sun spaces. That option is certainly the simplest for people who have little time to devote to the supervision of contract work, no interest in directing the project, or who simply want to hand the project over to a seasoned professional for a minimum of headaches and hassles.

But you will still have certain responsibilities, from the earliest planning stages through completion of the building process. Even if you're away from home during the day when the construction takes place, you should monitor the progress and try to meet with the contractor as often as possible, especially if problems arise. You may initiate a system of leaving messages back and forth to facilitate communication, or perhaps establish a routine of daily telephone calls. Maintain contact to reduce the chance of misunderstandings.

You may choose, instead, to act as your own general contractor, subcontracting the various jobs—the foundation, plumbing, electrical wiring, assembling the sun space, etc.— that the project requires. If this is your decision, you'll need to consider the project a full-time job. Be sure you have the time, energy, and inclination before you undertake it. If you

Courtesy Solar Resources, Inc.

Courtesy Solar Resources, Inc.

A do-it-yourself sun space requires patience, construction expertise, and a major commitment of time and energy. But, if you decide to forge ahead with such a project, you'll enjoy not only the resulting sun space, but the satisfaction that comes from successfully completing a big job.

feel that this is a good way to save money, be very sure you have not only the time but the expertise to see the project through to its completion. In the long run, you may decide that it's well worth it—for your peace of mind as well as your pocketbook—to hire professionals to handle the entire project.

Building the sun space, depending on the size and complexity of the project, may take anywhere from a week to a month—longer if delays are caused by weather or the unavailability of materials. The sooner you can get through your preliminary planning and research stages, the sooner you can move on to enjoying your sun space, for as long as you live in your home. That end result will make up for any inconvenience caused by the building process, which, if you view it as a necessary means to a desired end, will soon fade from memory.

Checklist For Contracting And Building Your Sun Space

• talk with friends and family for recommendations

• check references

• visit in-progress job sites

• visit completed job sites

• contact the Better Business Bureau, Consumer Affairs office

• take bids on project

• decide on contractor

• have contract drafted

• determine payment schedule

• have attorney examine contract before you sign it

• contact local building department regarding building codes

• request variance, if needed

• obtain building permit

• supervise building process

• expect delays and cost overruns

• communicate effectively with contractor

• relax and enjoy your sun space when it's finally done!

A hot tub built into a sun space—a delight to the senses and an undeniable luxury.

Courtesy Four Seasons Greenhouse

CHAPTER 4

Maintenance and Preventive Care

Despite the claims of manufacturers or dealers, no sun space can truly be considered maintenance-free. But a sun space need not require extensive maintenance, just a minimum of preventive care and a more thorough checkup in the spring and fall. With just a few hours of seasonal care your sun space will reward you with years of pleasure and enjoyment.

Preventive Care
Glazing

Regularly examine the glazing, both indoors and out, for damage, cracks, fogging, and leaks. Cracked or damaged glazing should be replaced promptly to prevent further damage to it and the framing materials in your sun space. Check with your dealer or manufacturer, consult your owner's manual or hire a professional to replace it.

Double-glazing that becomes fogged between panes indicates the presence of condensation. Replacement costs may be covered under your manufacturer's warranty; again, check with your dealer or manufacturer's representative to be sure.

Keeping the glazing clean may be more of a job than it sounds, especially if your sun space incorporates a large amount of overhead glazing. Most sun space owners are content to clean the overhead glazing themselves once or twice yearly; many feel more comfortable contracting the work. If you decide to do it yourself, remember to anchor the ladder firmly and never walk on the glazing surface. Use special care with aluminum-framed units, which can become extremely slippery when wet.

Use a cleaning solution of warm water and a small amount of ammonia, vinegar, or bleach. Customized squeegees that fit the panes exactly, soft brushes, and extension poles simplify the job; if you have to scrape the glazing to remove stubborn stains, take care to avoid scratching the surface. Because bright sunlight causes the water to evaporate quickly, resulting in streaks, schedule your cleaning session for early morning or late in the afternoon.

Maintenance

Left: Twice-yearly major cleanings, and occasional touch-ups and preventive maintenance, will keep glazing relatively clear and clean. Seasonal inspections of your sun space will head off any major problems.

Below: In a frequently used, relatively small sun space like this one you may want to clean the glazing regularly for maximum enjoyment and clear outdoor views.

Courtesy Pacific Coast Greenhouse Mfg. Co.

© Robert Perron

Interior

Pay regular attention to the condition of your sun space; examine it for leaks (both air and water), condensation, and cracks in the glazing. Again, check your owner's manual and check with your dealer; the necessary repairs may be under warranty.

Check the condition of any shades or blinds. Do you need greater insulation or more shading? Are the blinds or shades in good shape, or do they need replacing? If you have installed a shading system which uses tracks, the tracks may need to be lubricated (a shot of silicone usually does the trick).

Evaluate your day-to-day satisfaction with your sun space, and the functional effectiveness of the extras you've added to increase your enjoyment of the room. Some modifications may be necessary after you've lived with it for a while. Pay attention to the complaints of friends and family who use the room.

Courtesy Pacific Coast Greenhouse Mfg. Co.

Courtesy Green Mountain Homes

Above: Preventive maintenance will keep your sun space impervious to the weather—and keep you snug and cozy indoors with a lovely view of nature's show.

Left: Carefully examine the interior features of your sun space every spring and fall, both inside and out. Simple, natural furnishings help keep the task an easy one.

Exterior

It's usually easiest to inspect the exterior of the sun space when you're cleaning the glazing. Thoroughly examine the caulking, weather stripping, and glazing tape, and any screws, nails, or other fasteners that may loosen with changes in temperature throughout the year. Replace, repair, or tighten where it is appropriate.

In units utilizing wooden framing, carefully examine the

© Jerry Howard/Positive Images

Above: **Pre-winter care will get you through the coldest months safely and comfortably. For best results, remove any overhanging branches, check caulking and sealants, and in the winter, keep drifting or piling snow from accumulating.**

Right: **A sun space can be beautiful, not only from the inside looking out, but from the outside looking in. This attractive backyard hot tub has a lovely view of a cozy, sparkling sun space.**

condition of the wood; it may need refinishing or treatment with a wood preservative if it has become weathered. In aluminum-framed units, check for clear drainage of weep holes, intact gaskets, and bar caps to maintain a weather-tight sun space.

Before winter begins, remove any overhanging branches that could break loose during a severe storm. You don't want to expose yourself and your family to the danger of a branch crashing through the glazing of your

sun space. In snowy areas, be sure to remove drifts and piles of snow that may accumulate on or against your sun space. The weight of the snow, along with its freezing and re-freezing, can cause seepage, wood rot, and damage to both the glazing and the framing materials.

Following these common-sense maintenance tips will keep your sun space clean, comfortable, and nearly care-free—and can help prevent problems later caused by the lack of proper maintenance.

Maintenance Checklist

• Keep glazing clean, inside and out

• Replace cracked or fogged glazing

• Replace loose or dried-out caulking, weatherstripping, and rubber gaskets with compatible materials

• Check for air or water leaks

• Examine the condition of the framing: Wood may need refinishing; anodized or painted aluminum may require touch up paint

• Remove dead tree limbs and branches

• Inspect condition of shades, awnings, blinds; reevaluate their effectiveness

• Examine ventilation ducts and fans; keep free of dust, animal nests

• Keep floor clean; replace tile grout if needed

Trouble-Shooting Maintenance Problems

Is your sun space too hot?

• Close off from the rest of the house

• Add external shading (blinds, awning, deciduous trees)

• Add internal shading (blinds, curtains, shades)

• Improve ventilation throughout sun space with the use of fans, vents, open windows, and doors

Is your sun space too cold?

• Consider insulating shades

• Use a portable electric heater

• Install a wood-burning stove

• Add heat storage materials

© E. Alan McGee/FPG International

Left: This unusual design, with its high ceiling and strong beams, combines the elegance of wide expanses of glass and the solidity of wood. Both are simple, classic materials long used in the design of sun spaces.

Courtesy Four Seasons Greenhouse/System 4

Above: **Remove piles of drifting snow as soon as possible to prevent maintenance problems. Water seepage, sealant failure, and wood rot can occur without proper attention.**

Is there an excessive glare in your sun space?

• Add shade with blinds, awnings, large plants, and shutters

• Minimize the use of white framing materials and accessories

Is condensation developing between layers of double glazing?

• Consult the manufacturer for warranty information

• Replace affected panels

Is condensation developing on the inside of the glazing?

• Improve the air circulation

• Install a vapor-barrier cover over pool, hot tub, or spa

• Dehumidifier may be needed

Is there insect infestation among the plants?

• Check all plants carefully before bringing them into the sun space

• Discard infested plants

• Check with local nursery for natural alternatives to chemical pesticides

• Grow plants from seeds

A STEP-BY-STEP GUIDE
How to Maintain Your Sun Space

Step 1: Glazing

Step 2: Caulking,
weatherstripping, and gaskets:

Step 3: Interior

Step 4: Exterior

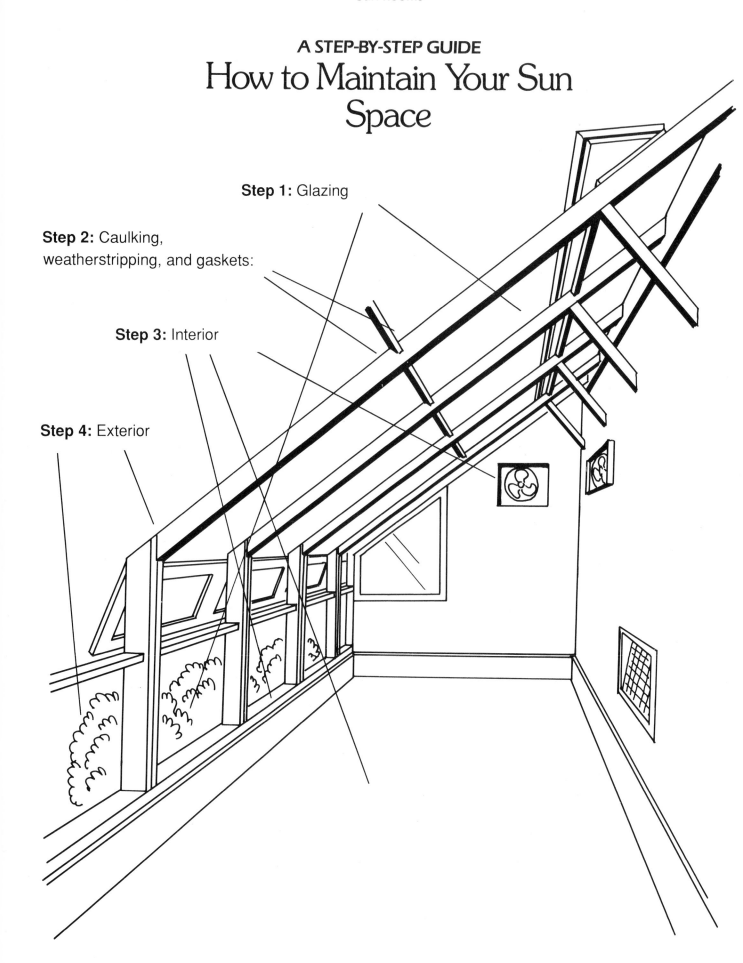

Every spring and fall the following procedures should be conducted to maintain your sun space.

Step 1: *Glazing:*

- Keep glass clean by washing it each spring and fall
- Use specially shaped squeegees for extra ease, or contract the work to someone else
- Replace any cracked glazing
- Replace double-glazing that has become fogged (indicative of a broken seal)

Step 2: *Caulking, weatherstripping, and gaskets:*

- Examine all weatherstripping, caulking, and gaskets for deterioration; replace or recaulk as necessary

Step 3: *Interior:*

- Examine the wooden framing and sills for dampness, softness, and wood rot; replace if necessary
- Check all the doors and windows for air-tightness; add or replace weatherstripping as necessary to maintain air-tight unit
- Examine the shades for wear, the tracks for obstructions and smoothness of operation, repair or replace the shades as necessary; lubricate tracks if needed
- Clean exhaust fans, heat exchange fans, air vents; replace air filters
- Examine flooring for wear; replace tiles, worn carpet, or other materials as needed
- Examine plants for insect infestations, excessive growth, or crowding; rearrange, treat, or replace as necessary
- Check furnishings for color-fastness, wear; rearrange or replace as desired

Step 4: *Exterior:*

- Remove any overhanging branches or overgrown trees or shrubs
- Examine weep holes in framing; remove any twigs or materials that can clog or interfere with proper draining
- Examine framing for excessive weathering (especially wood); apply appropriate preservative
- Use touch up paint on any chips, scratches, or nicks in the surface of the aluminum framing
- Examine the outdoor environment for scenic value; prune, cut back, and replant to improve the view from indoors.

Courtesy Laura Ashley

CHAPTER 5

Plants: The Greening of Your Sun Space

There's something very special about bringing plants into a living area. The touch of nature they provide gives any room a feeling of lightness, of space, and of the beautiful outdoors. Nearly every home has a few plants, ranging in states of health from straggling to robust, sharing living space with its occupants. Some individuals pride themselves on their "green thumbs"; others confess that they have no idea how to care for even the hardiest green plants—and they are afraid to try.

There's really no reason to fear the care of plants; it's like any other endeavor—fairly straightforward and simple, once you understand the basic principles. A little trial and error may be needed to give you the confidence and the experience you need to devote a substantial amount of time and money to plant care. Chances are, if you're attracted to the idea of a sun space, plants are already important to you and you may have quite a bit of expertise in basic plant care. If so, the following section will serve as a simple review; if not, here are the basics to get you started.

Basic Plant Care

Plants have few real needs—perhaps that's one of the reasons we tend to gravitate toward them. Even a minimum of care can gain fantastically beautiful results as your plants reward your efforts by growing strong, healthy, green, and—especially in a sun space—noticeably larger in a short period of time.

Photosynthesis, the chemical process by which plants convert solar energy into carbohydrates, is an extremely complex process. But when we care for plants, all we have to do, really, is to create the optimum conditions for photosynthesis—that essential life process—to take place. The process requires the right amounts of light, water, fresh air, proper temperatures, humidity, and soil nutrients. When you choose to take plants indoors, it's solely up to you to control conditions in your plants' environment so that their rather modest requirements are met. Nowhere in your home is that simpler than in the sun space designed to house greenery.

First consider the basics required for plant life.

Left: It doesn't take a ''green thumb'' to grow lush plants like these. A regular commitment of time, a little information, and some practice are all that is required.

Below: People and plants can be quite compatible when greenery is properly selected and carefully tended.

© Derek Fell

Light

The most common problem most people have in growing house plants is not enough light in their rooms. That's not normally a problem in a sun space. In fact, you may find that your sun space admits too much light that must be filtered—with shades, blinds, awnings, curtains, or other sun blocks—for optimum plant growth. But to be certain of the correct light levels in a sun space designed for plants, consider the use of overhead glazing, and be sure that any plant beds are positioned parallel to the bottom of the vertical glazing.

The amount of light your sun space receives will determine the type of plants that will grow successfully in it. Flowers will bloom, and fruits and vegetables will grow best in a southern-facing sun space that gets maximum year-round sunlight. Tropical plants, which flourish in filtered light, will grow best in sun spaces facing either east or west, and those that are somewhat shaded.

Water

Because of the abundance of humidity in most sun spaces, plants living in them require somewhat less water than plants that grow outdoors, or even those that grow in other rooms of the house. Indeed, over- and under-watering tend to be the two greatest problems in house-plant care.

Courtesy Jardin Inc./photo: © Bob Emmott

© Jerry Howard/Positive Images

Left, top: Lots of light, good ventilation, proper humidity, and the right temperatures and—voila!—lovely, healthy green plants.

Left, bottom: The right amount of water—not too little or too much—is essential for proper plant care. This greenhouse is obviously carefully tended.

Would-be gardeners are either overprotective and water their plants far too often, or they allow their plants to dry out to a point where they're parched and their soil becomes rock-hard. Naturally, neither extreme is conducive to healthy green growth. The primary key to watering is to be thorough—to saturate the soil to the point where water leaks out the drainage hole in the bottom of the pot. Only when the soil is saturated can you be sure that the plant's root system has received adequate moisture. The secondary key to watering, for most plants, is to allow the soil to dry out between waterings to prevent overwatering. Resist the impulse to offer your plants a drink every day. Their roots—indeed the entire plant system—will grow healthier and stronger if they are not waterlogged.

© Jerry Howard/Positive Images

Right: Examine your plants regularly to keep them healthy and free of disease or insect infestation.

Fresh Air

Good circulation and air flow is important to plants, as it is to people. Without adequate air flow, humid air clings to the plant leaves and can lead to sogginess and leaf rot. Your sun space should be adequately ventilated, not only to provide fresh air for the plants but also to circulate the oxygen they produce. Fans, vents along the sun space at the kneewall and roof line, and opening windows and doors will all facilitate the movement of air through the sun space.

Temperature

Plants thrive in the same temperatures comfortable to humans, primarily between 60 and 85 degrees F (15 to 29 degrees C). Too much heat or cold can kill off sensitive plant tissues, so keep the temperatures well within the tolerance limits of your plants, remembering that some are far more sensitive than others. If the nighttime temperatures dip below comfortable levels, or if the daytime temperatures soar above them, you may need to use shades, shutters, or blinds, or auxiliary heating or cooling units to protect your plants.

Another caution: Do not allow plant stems, leaves, or branches to come in contact with glazing surfaces that may conduct heat or cold, thus damaging plant tissues.

Humidity

Sun spaces are natural humidifiers, as is the presence of a number of plants concentrated in a small space. You can increase the level of humidity around your plants by grouping them close together

in plant benches that are lined with pebbles. If you keep the pebbles moist, the evaporation of the water will provide humidity levels your plants will love. Another way to increase humidity is to mist the plants occasionally with a water sprayer.

The problem of too much humidity can arise, especially during spells of humid, overcast weather conditions. Condensation on the glazing surfaces, stuffy, stagnant air, and the increased likelihood of fungal diseases taking hold are a few of the consequences of too-humid conditions in the sun space. Avoid the problem by watering only when necessary, and when it does become too humid, vent the sun space to dry it out a bit.

Soil

The simplest solution to the proper soil mix is to purchase premixed commercial soils. Commercial preparations are formulated specifically for many different types of plants, which can eliminate much guesswork. Because premixed soils are sterilized, they cannot harbor potentially debilitating plant fungus or diseases, which could affect the entire ecosphere of the sun space.

As you become more knowledgeable about your plants' needs, you might want to modify commercial soils with a little more sand here, a little more gravel there, to create a soil that you—and your plants—prefer.

© Jerry Howard/Positive Images

Above: These healthy specimens can only grow and flourish under optimum conditions. Since a greenhouse is an artificial environment, it's up to you to provide the right conditions for the plants you choose.

Left: Those who love plants may want to construct a full greenhouse to get the most joy from the plants. Many attest to the therapeutic value of getting a little dirt under the fingernails; remember, though, a full greenhouse does require a commitment of time and energy to keep plants healthy.

Nutrients

The three essential nutrients for plant growth are nitrogen, phosphorus, and potassium. Nutrients are absorbed by plants' root systems, and, especially in potted plants, the soil is eventually depleted of them. To replenish the vital nutrients, you must choose and properly administer plant foods and water-soluble fertilizers. The natural temptation when feeding plants is to think that if two drops are good, four would be even better. That's not the case. Overfeeding your plants can burn the roots and damage their health. Slow-growing plants may need to be fed supplemental food only two or three times a year, while rapid growing ones may need to be fed whenever a new leaf is produced. Check with your local greenhouse, nursery or a gardener for expert guidance.

There are no complicated secrets to plant care, but no one would deny that some people have more of an affinity for caring for greenery than others. That "green thumb" probably comes from learning the basics of plant care and getting to know the plants, both as individuals and as species. It takes some time to feel comfortable with the plants you bring into your home; most people feel it's worth it.

Caring for plants, especially a large number of them, can be an awesome, time-consuming task. In addition to the aforementioned basics of regular watering, feeding, and misting, plants really need a certain amount of maintenance for optimum growth. Regular attention in the form of removing dead leaves, pruning and cutting back, inspecting foliage for fungus or infestations, and keeping plants clean can take up hours each week. Most people feel the time involved is therapeutic and enjoyable; indeed it should be. If plant care becomes a chore, you need to reduce the number of plants or get some help.

Above: **This greenhouse/sun space is quite compatible with the kitchen and living area of this home.**

Left: **Displaying your plants attractively is part of the fun. Hunt down old bits of crockery, pots, and special flats that appeal to you.**

The Sun Space As A Planting Space

Puttering around with your plants in your sun space can be a real pleasure, especially if you've designed the room as a functional and beautiful place. The plants do require some regular attention, but the structure itself does not require much more maintenance than any other sun space. To keep the unit clean and fresh, you should thoroughly clean and scrub your sun space once a year with a solution of 10 percent bleach, 90 percent water. At the same time, you should scrub the mineral encrustations from your clay pots, and give the entire plant area a general once-over, noting anything that needs further maintenance or attention.

Your early selection of materials and layout will have a lot to do with your future happiness with it. For example, consider the floor—it should be as low-maintenance as possible. For sun spaces used solely as garden rooms, consider using a brick, concrete, or slate floor.

The natural materials are not only attractive—they cannot be damaged by the inevitable soil spills or water drips. Many greenhouses are also designed with a drainage system in the floor for simple cleanup.

Position your plant beds and benches where they will absorb enough light, yet still be convenient for you to tend. You don't want to be stooped uncomfortably over your plants, nor do you want to reach high above your head to gain access to them.

Naturally, your greenhouse should have a sink and convenient water supply; a work space for repotting, pruning, and cleaning plants; shelf and storage space for your tools and supplies; and at least a chair or two where you can sit, relax, and enjoy the fruits of your labor.

Selecting And Displaying Plants

As you select the plants for your sun space, you must consider the physical limitations of the unit: the amount of light, the actual size of the room, and whether or not the particular plant could be expected to flourish in your sun space environment. You should also determine exactly how those plants will be displayed. Will they be hanging plants, suspended from the ceiling, or hanging from brackets or hooks attached to the walls or framing materials?

© Jerry Howard/Positive Images

Make sure that the suspension devices are anchored sturdily enough to support the weight of the hanging plants. Plant shelves, benches, and beds should be built to your specifications, finished to match the overall feel of your sun space, and carefully lined with metal or plastic to protect the wood.

Many sun space owners have been disappointed when they begin furnishing their sun spaces with plants. A few small plants simply look lost in any room; but don't expect too much too soon. It takes time to develop the lush, green growth and almost-overgrown look that you probably desire. As you purchase your plants, keep in mind the need for a sense of balance between large and small plants. You may opt for the drama and impact of a few—or several—large plants complemented by smaller ones (which will surely grow quite rapidly in the optimal

© Nancy Hill / Courtesy House Beautiful's Home Remodeling & Decorating

conditions of the sun space). Larger plants are also hardier, and therefore simpler to grow, than seedlings or smaller, younger plants.

A well-planned sun space takes not only the sizes of plants into consideration, but also their variety in terms of shape, texture, and color. A beautiful, healthy assortment of greenery, carefully arranged for their complementary and compatible characteristics, is the goal in selecting plants for your sun space.

The above discussion presupposes that plants are chosen primarily for their beauty, appropriate size, ease of care, and general appeal. There are additional criteria to consider, particularly if you have a specific interest you want to pursue with your plants. For our purposes, those interests can be divided into two major types of plants: ornamental (flowering and tropical) and food bearing.

Above: **Plants are always graceful, attractive, and welcome additions to the open areas in a sun space.**

Opposite page: **Plants should be placed at comfortable levels to allow for easy maintenance. Wooden-planked flooring simplifies clean-up.**

Ornamental Plants

When the beautiful, colorful blossoms of flowering plants or the perfectly formed new leaves of green plants bloom in the sun space environment, you can gain a sense of pride and connection with the natural world. The richly colored, intricately formed foliage and flowers are your tangible reward for hours of careful attention to your plants. Raising your own greenhouse flowers can be a heady experience that brings you true delight.

Greenhouses are usually separated into two or three different categories based on their temperature ranges: warm, intermediate, and cool. A warm greenhouse is one where the daytime temperature ranges from 65 to 75 degrees F (18 to 24 degrees C), and the nighttime temperature from 60 to 70 degrees F (16 to 21 degrees C). Intermediate

greenhouses have daytime temperatures ranging from 60 to 70 degrees F, and winter nighttime temperatures of 50 to 60 degrees F (10 to 16 degrees C). A cool room's daytime temperatures range between 50 and 60 degrees F, and may dip down to 45 to 50 degrees F (7 to 10 degrees C) at night. Naturally, the different greenhouse environments will determine the types of plants that grow in them.

In general, many plants and people can coexist fairly comfortably in living spaces that have the temperature characteristics of cool or intermediate greenhouses—at least during the day. The cooler nighttime temperatures preferred by many plants are uncomfortably chilly for most people. But, if you opt for a cool or intermediate room, it's fairly simple to close the cool sun space off from the rest of the home with an insulated sliding door. As you select your plants, be certain you've determined the temperature range of your sun space and plan accordingly.

Warm rooms

Flowering plants that can withstand the warm temperatures of sun spaces also designed as living areas include: delicate fuchsias, colorful begonias and cyclamen, hardy geraniums and kalanchoes, and nearly everyone's favorite, African violets. Tropical green plants that typically share living space in many homes can grow spectacularly in warm sun spaces. Tough,

hardy, and easy-to-grow tropicals include the familiar spider plants, ivy, several palms, and philodendron. More exotic tropicals include the huge papyrus, the lovely white arum lily, and the brightly colored anthurium, bougainvillaea, and coleus.

Intermediate rooms

Perhaps more plants favor intermediate temperatures than any others. The wide range of these plants includes everything from the lovely Norfolk Island pine to exotic orchids, from the dramatic bird-of-paradise to romantic roses. Others include many varieties of ferns, especially asparagus, maidenhair, staghorn, and the

Above: **A multi-level approach: sun space below, skylights above. Just the thing for beautiful living areas full of light and air.**

Opposite page: **The time invested in caring for plants is richly rewarded with healthy, green foliage.**

© Robert Perron

Left: Greenery inside and out enhances the home's environment and makes it a nicer place in which to live.

© Jerry Howard/Positive Images

Below: **The bright blooms of spring-time bulbs are a lovely reminder of the wonders of nature's cycles. The light of a sun space allows these beauties to bloom indoors.**

Victorian conservatory favorite, the Boston fern. African lilies, begonias, fuchsias, spicy-scented geraniums, and jasmine all favor intermediate temperatures.

Cool rooms

Rooms that are uncomfortably cool for people are perfectly suited for a number of beautiful blooming plants. These include delicate camellias, beautiful azaleas, standard cymbidium orchids, primroses, fragrant gardenia and freesias, and long-lasting cape cowslip —all of which flourish in cooler-temperature sun rooms.

As you gain experience and learn more about plant care, you may want to try your luck with forcing various bulbs for springtime blossoms, perhaps a cactus garden for a starkly attractive desert feel, or starting delicate flowering annuals for eventual planting outdoors. These varying challenges will add an exciting new dimension to the enjoyment of your sun space.

Ornamental plants can be displayed in any number of lovely ways; many look best when hung suspended from hooks in the ceiling, walls, or along the framing bars. Use a variety of hangers and pots and keep in mind the need to vary texture, color, and height for added interest in your sun space. Use your imagination when you consider what to use as plant stands. Certainly you may use specially crafted tiered plant stands or jardinieres, but consider what else

Courtesy Laura Ashley

Above: The traditional beauty of this sun space—all glass and light—is delicate, airy, and very Victorian.

Opposite page: The lush growth of these plants is facilitated by a number of factors, including excellent ventilation provided by windows that can open.

you might adapt to suit your tastes and add visual appeal. Some suggestions include Baker's racks, old Victorian fern holders, stacked glass bricks for a modern look, and industrial shelving for a high-tech look.

Plants grown in sun spaces are simplest to tend when they're grown in pots. They can be easily moved and maintained in containers that can be as basic as clay pots, or as ornate as brass planters. Collect interesting crocks, ceramic pots, old tubs, and even a horse trough for varied accents in your sun space.

Food Plants

You don't really need a huge garden in order to grow your own food. There are several fairly easy "crops" you can grow in a sun space; indeed the first conservatories were referred to as "orangeries," for the miniature oranges that were grown in them. Dwarf citrus trees, including orange, lemon, lime, grapefruit, and tangerine, all do quite well in sunny, intermediate, or cool sun rooms. Strawberries are fairly simple to grow in large pots or half-barrels; grapevines can be trained to grow up trellises, providing not only fruit but also a romantic shade canopy for the sun space. Other fruits that can be cultivated in sunny greenhouses include melons, figs, peaches, and nectarines.

Vegetables, too, can flourish in sunny, cool, or intermediate sun spaces. Some of the

simplest to grow include radishes, eggplant, various members of the squash family, peppers, lettuce, carrots, and tomatoes. Fresh greens in the winter can add a delightful touch of springtime to your dining room—Swiss chard, Chinese greens, some beans, peppers, and spinach can be harvested in winter months.

Fresh herbs add wonderful flavors to your food and beauty and fragrance to the sun space, making them big favorites for indoor cultivation. Herbs generally produce their

best flavors when they're grown slowly in full sunlight. Popular, easy-to-grow herbs include parsley, oregano, chives, thyme, rosemary, basil, dill, mint, and garlic. Most herbs are fairly hardy and therefore rewarding to cultivate, but they are—like most plants—susceptible to damage from cold.

There is nothing like the experience of walking into your own garden to pick lettuce and tomatoes for a salad, dill to season a sauce, eggplant for a casserole, or strawberries

for dessert. The freshness, color, taste, and fragrance simply cannot be duplicated in store-bought food; the fact that the fruit and vegetables were carefully cultivated and nurtured by your own hand adds another dimension to the dining experience. Growing food in your sun space makes it possible to dine on fresh, home-grown produce all year round.

Insect Infestations

As idyllic as it sounds to grow your own lovely plants, flowers, and food indoors, there are a few pitfalls to avoid, most notably the infestation of your plants by unwelcome insects. Whiteflies are the most common infestation; others include spider mites, mealybugs, aphids, scale, and various gnats. A sun space overrun by such intruders can be difficult to restore to order; steps taken to prevent infestation are well worth the effort. Good ventilation, keeping the sun space clean, using sterilized soil and clean pots, and isolating new or infested plants are all effective measures for keeping the sun space clean and free of pests that have no place in your home.

If insects do take hold in a sun space, particularly one used as a living area, the use of natural insecticides is definitely preferable to the use of toxic chemicals. The simplest

Courtesy Solite Solar Greenhouse Corp.

solution is to inspect and clean your plants regularly, examining the undersides of leaves and the condition of the soil, and becoming aware of the overall health of your plants. Water and insecticidal soaps should be used for cleaning foliage; rubbing alcohol swabbed or sprayed onto leaves will take care of many infestations, including whiteflies. Serious, uncontrollable infestations may require the use of chemicals; check with your local nursery for recommendations and advice.

Beautiful, healthy plants will add drama, a light feeling of airiness, and a sense of nature to your sun space. Whether you bring plants into your living area or use a sun space solely for the cultivation and propagation of botanical beauties, you're sure to be rewarded with many hours of enjoyment for your efforts.

© Jonathan A. Meyers/FPG International

Above: A sun space can even brighten a basement, allowing for additional living space and a place where plants can grow at—or below—ground level.

Left: A lovely room addition adds much to this home, but hardly reduces the size of the backyard at all.

Right: What better place to enjoy a hot tub than in a climate-controlled sun space that provides privacy, a view, and lets you laugh at the weather.

© Jerry Howard/Positive Images

A STEP-BY-STEP GUIDE
How to Build a Plant Bench

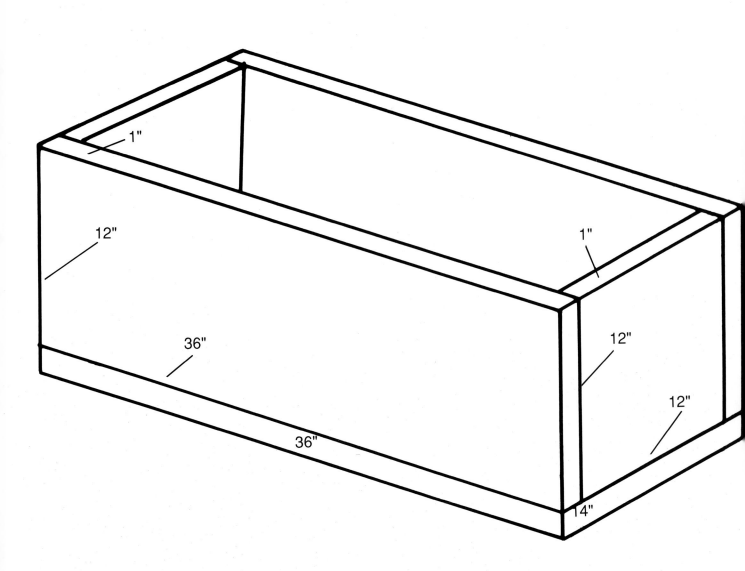

When building a plant bench, there are a few important considerations to keep in mind. The height of the bench should allow you to comfortably tend the plants without having to bend over too far, and the depth should allow you to easily tend all of the plants on the bench without having to stretch out or reach back too far.

Step 1: For this bench, you will use 1-inch (2.5-centimeter) thick cedar or redwood throughout: two pieces of cedar measuring 12x36 (30x90 centimeters) for the front and back, one piece measuring 14x36 inches (35x90 centimeters)

for the bottom, and two more pieces for the sides, measuring approximately 12x12 (30x30 centimeters). The exact size will be determined later.

Step 2: From underneath, nail the bottom piece of wood (14x36 inches) onto the front and back pieces (the 12x36 pieces) using long, thin galvanized or cement-coated nails.

Step 3: Measure the exact distance between the front and the back and cut the remaining pieces (they should measure around 12x12 [30x30]) of wood. Insert these two pieces flush with the ends and nail from the front, back, and bottom, every 3 inches (7.5 centimeters).

Step 4: For drainage, drill 4, ¾-inch (1.875-centimeter) holes in the bottom. If you do not drill holes in the bottom, add 1 to 2 inches (2.5 to 5 centimeters) of gravel or rocks to the bottom of the bench.

CHAPTER 6

Luxurious Extras and Finishing Touches

Now comes the time for fun. You have gone through all the steps to ensure that your sun space is properly situated and built to serve your needs. Once it's finally built, you can add all the luxurious extras and finishing touches that will turn the sun space into the room you've dreamed about for so long.

Consider now what first attracted you to the idea of building your own sun space. Keep in mind the charming little details that you want to include—details that can be as simple as hanging prisms, crystals, and wind chimes, or as complex as an intricately patterned tiled floor.

You may want to add the finishing touches slowly, but this chapter will provide you with some ideas that will make your sun space exactly what you want. Let your imagination soar as you consider your options, and design the room that most creatively and accurately reflects your tastes and interests.

Luxurious Extras

There are many major investments and projects that you might want to add to your sun space, for comfort or just outright luxury. These include such extras as a built-in bar, lap-swimming pool, hot tub, or spa; and decorative yet functional items, including

shades, blinds, thermal curtains, and ceiling fans, as well as other furnishings that are sturdy enough to withstand the sunlight and heat of a sun space.

Built-in Bar

Will you be using your sun space for entertaining regularly? If so, you might want to consider installing a built-in bar unit, complete with refrigerator, sink, and shelves to hold glasses and bottles. The unit should be situated well away from the glazing of the sun space, preferably against or next to a solid masonry wall in the room. Naturally a bar

unit works best in a large sun space designed to be used primarily as a living area.

Lap-swimming Pools

In this age of physical-fitness consciousness most of us have begun to realize that one of the best things you can do for yourself is to get regular exercise. Swimming is by far one of the most popular participatory sports in America—but most people consider it a seasonal activity, unless they have year-round access to an indoor pool. What better way to get your daily exercise than to take a swim in your own swimming pool, installed in

Courtesy Four Seasons Greenhouse/System 4

Courtesy Solite Solar Greenhouse Corp.

your sun space? Lap-swimming pools, usually fairly narrow, shallow pools, can run the length of the sun space, providing the means for you to improve your health and adding a beautiful touch to your room as well.

Hot Tubs and Spas

Humans seem to have a special affinity for water, and nowhere is that more obvious than in the increasingly popular use of hot tubs and spas. These beautiful, undeniably luxurious additions to a home can become the central gathering place for the entire family. Somehow, conversations

can be more pleasant, open, and forthcoming when they take place in the comfort of a hot tub or spa. A sun space is the perfect shelter for a hot tub; it provides protection from the elements, a beautiful view, and a sense of intimacy.

Installing a hot tub or spa in a sun space requires not only advance planning but a few additional precautions in the initial building stages. A hot tub, filled with hundreds of gallons of water, can weigh more than a ton. Make sure the foundation is built to take the weight. Do not place the hot tub in the middle of the sun space; the concentrated weight in the center of the

Above left: **An enclosed pool is especially desirable in cold-weather climates.**

Above: **Physical fitness can start at home with a lap pool enclosed for privacy, easy maintenance, and year-round use. No need to deal with crowded public pools when you own your own.**

Courtesy Four Seasons Greenhouse/System 4

floor can cause shifting or buckling, and you don't need that. Instead, place the unit in the most pleasing corner, remembering the need for access to the plumbing source. To alleviate the problem of excessive chemical-laden humidity, heat, and evaporation, which can corrode the aluminum frame of your sun space, a well-fitting, heat-retaining cover should be kept on the unit. If high humidity becomes a problem, improve the ventilation by installing vented windows, fans, or a dehumidifier in the sun space.

Other Options

Although a sun space is undeniably a luxury room, it can certainly serve a utilitarian purpose. Consider utilizing a sun space as a work room, an office, or an exercise room.

Work Rooms

Hobbyists who enjoy various activities from needlework to stained glass, from geneaology to stamp collecting, always appreciate having their own room where they can spread out and enjoy their projects. Working in a simply furnished sun space can enhance the enjoyment of any hobby. Equipped with a good work surface, shelves where needed, and proper lighting, a sun space/work room can be a delightful addition to the home of even the most casual hobbyist.

Office Space

Those who work at home, especially those in the creative professions—including writers, artists, architects, and photographers—face the age-old problem of creating an office that is at the same time functional, private, attractive, and removed from the rest of the home to provide the peace and quiet necessary for productive work. What better solution than the properly placed sun space? Ideally, an office would be positioned away from the highly trafficked parts of the home, and with a beautiful view (natural or created with plants) for a creative environment and a wonderful alternative to four walls and a window.

Exercise Room

Everyone wants to keep in shape, look good, and feel even better. All too often, health-conscious, busy people join health clubs or gyms, planning to work out every day, but those good intentions fade with the stresses, strains, and time demands of daily life. But a hectic schedule is not a good reason to keep from working out.

Why not use a sun space as an exercise room? It's simpler than ever today to equip an exercise room with a stationary bike, a padded surface for calisthenics and floor exercises, a video monitor to help you work out to the latest aerobic videotape, a great sound system to help you stay with the program, a weight bench and weights, a ballet barre on one wall, a full-length mirror—add a few posters of your favorite fitness advocate, and you're ready to get in great shape. Your exercise room/sun space could be much more enjoyable than any health club; it's customized to suit your needs exactly, and it never closes!

Naturally, when planning a sun space for a specific purpose like an exercise room, office, or work room, you'll need to consider special requirements. The exercise room must have proper ventilation; the office or work area requires proper shading and perhaps air conditioning in a warm climate, heating in a cold one. Consult your builder or contractor for guidance.

Opposite page: **Pretty place—inside and out—that's what sun spaces are all about. Here, wicker inside and lovely trees outside combine for universal appeal.**

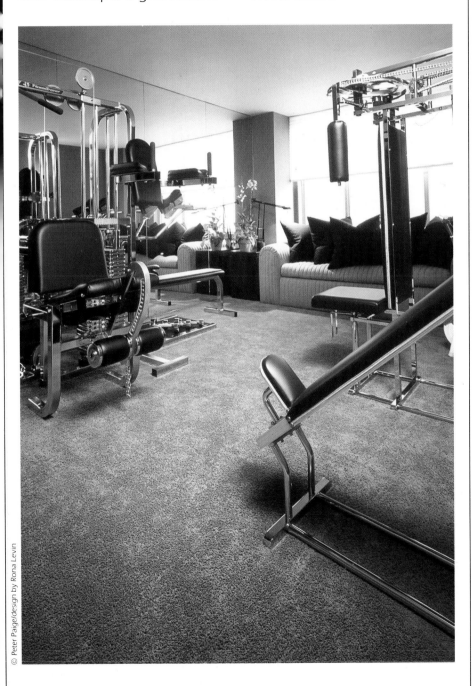

Left: **Why not work out in a sun space? Avoid lines, crowds, deafening tunes, and inconvenient hours by creating the workout space that works for you.**

© Peter Paige/design by Rona Levin

Opposite page: **The perfect garden setting—in the home and in the backyard—captures the beauty of the natural world.**

Below: **Shading systems are especially important in sunny areas. They help diffuse the light and shade you from the heat that enters the sun space.**

Shades

The major problem many sun space owners face is not concern over not enough sunshine or heat, but rather too much. Nearly every sun space, even those in areas with high amounts of solar radiation such as the Southwest, can be made more comfortable by using shades to deflect excessive amounts of sunlight.

Many manufacturers have devised shades specifically for use in sun spaces; some have developed extremely sophisticated—and expensive—designs that incorporate the use of thermostats, rolling tracks, timers, remote devices, and automatic electric controls. Others have developed simpler manual shades that roll on tracks that you raise and lower as you desire. While manual designs require you to raise and lower shades as the sun moves across the sky, they are much less expensive than automatic ones.

When you consider shades, remember that along with blocking the sunlight, they also effectively block the view and much of the natural light, presumably the primary reasons you originally constructed the sun space. Also remember that although shades reflect sunlight, they do not effectively reduce all of the heat from the sun. Quilted, insulated shades and thermal curtains, however, not only substantially reduce heat gain in the summer, but prevent heat loss in the winter. They are extremely expensive accessories, but may be worth it to you in terms of energy savings in both winter and summer. Other comparable shades include insulated panels and shutters.

Other options that have little insulating value but function as sunlight diffusers include awnings, matchstick blinds, rice-paper shades, aluminum mini-blinds, and lacy curtains. Some dealers feel

Courtesy Comfortex Corp./Duette Smart Shade

Courtesy Paeco Inc./Sol*Area

that with the proper selection of insulating and reflective glazing, shades aren't needed except for privacy. Although high-tech glazings are definitely an asset, sun space owners would be wise to consider some sort of shading system, particularly for units that receive direct sunlight.

Ventilation Systems

Every sun space needs some form of ventilation, both to cool the room and to circulate fresh air for comfort and good health for plants and people. Natural ventilation can be built into the sun space during construction by incorporating inlet vents and windows built near the base of the structure, and outlet vents built high on the unit. By properly opening and closing vents you can effectively use natural breezes to cool and ventilate the sun space. Naturally, many sun spaces require more than just this passive system; ventilation can be increased with the use of exhaust and even decorative ceiling fans. Large, slow-moving fans are preferable to noisier, less effective high-speed fans. You may decide to choose a multi-speed fan or a thermostatically controlled vent fan that eliminates the need to monitor the temperature in your sun space. Sun spaces with pools or hot tubs require greater ventilation to prevent too-humid conditions; local building codes have specifications that must be met when using ventilation.

© Maria Pape

Above: **This comfortable room combines some of the most pleasurable aspects of a sun space: natural fabrics and materials, lots of light, greenery, and the openness of large expanses of glass.**

Furnishing Your Sun Space

You may know you want to use your sun space as a living area, but perhaps you're unsure about how to furnish it for comfort and practicality. We've already discussed materials for your sun space—natural wicker, cane, and willow, heavy-duty wrought-iron, and protected woods are best. But think now about specifics—how about a lovely wood-and-wrought-iron park bench carefully situated in front of a beautiful view? Maybe you prefer the light and airy look of wicker and willow. Practicality, ease of care, durability, and lightness of materials should be your primary concerns when you choose the furnishings for your sun space.

Consider the style of the rest of your home. Your tastes may run to sleek contemporary, traditional, country, or an eclectic mix of all of the above.

Every style can be adapted in your sun space with utterly charming results. Whether you prefer sleek Bauhaus chairs, Victorian wicker rockers, or tropical rattan, each can be used successfully. Simplicity is the guideline here; you want to avoid a sense of clutter. Opt instead for a lot of light and space, which will allow an airy feeling to come to the fore.

Think of the impact of color and texture—smooth summer-brights, icy-cool pastels, dramatic jungle prints, nubby, earth-toned neutrals, sophisticated, deep jewellike shades— when you consider fabrics and accents for your sun space. A cool, natural-colored sofa, white wicker divan, or wooden park bench can be brightened and warmed with a few plump pillows carefully chosen to convey just the right feeling. Simple wicker chairs can be dramatized with colorful chair pads splashed with bright floral prints.

As you furnish your sun space, carefully consider the limitations of the room. Don't expect to be able to keep books in it; the high humidity and heat will wreak havoc with your treasured volumes. Keep a few magazines in wicker baskets in your sun space instead. Framed prints, posters, and the like should also be avoided; opt for plants and other natural materials to decorate the walls simply and compatibly with the conditions of a sun space.

Display your collections of

© Bill Rothschild/design by Marilyn Rose

Left: A very traditional, very elegant approach to a sun space— complete with treasured accent pieces and carefully collected items.

natural materials—shells, rocks, geodes, tiles, glassware, crockery, statuary and ceramic pieces—in your sun space. They add those special touches of warmth, charm, and individuality and reflect your interests in the colorful world around you.

A touch of whimsy and a touch of color are always welcome in an informal sun space. In an aluminum unit, consider painting the framing bars a bright color to add warmth and splash; hang stained-glass panels, colorful wind catchers, and beautifully toned wind chimes for a light feel. Hang crystal sun catchers to send rainbows bouncing around the room; install mini-lights on the glazing bars for extra sparkle and drama. Why not consider a circulating fountain system, a small pond, or even an aviary for exotic birds? The list is endless.

The whole point of creating a sun space is to design your own version of an exquisite Victorian conservatory. Enjoy the sunlight, the view of the great outdoors, the cool of the evening, the warmth of the afternoon—it's all enhanced when viewed from the magical perspective of a sun space.

Right: **Wicker furniture, plants, plenty of sunlight—the quintessential sun space. This space bridges the gap between inside and outside with doors that slide open to provide easy access to the patio.**

Below: **The beauty of a modern-day conservatory. Well-integrated into the design of the rest of the home, complemented by outdoor features and furnishings, and better designed than ever before.**

Glossary,
Sources,
and
Index

Glossary

Annealed glass. Single panes of glass used in windows; annealed glass shatters into shards when it breaks, making it unacceptable for use in doors or overhead panels.

Conduction. The transmission of heat or cold from one solid object to another.

Convection. The movement of heat through fluids, including air.

Double-glazing. Two layers of glass in a sealed unit, with insulating air space between them; the most commonly used glazing system in most sun spaces.

Glazing. The transparent or opaque materials, such as glass, acrylic, or plastic, used between the framing bars in a sun space.

Greenhouse effect. Warming caused by the conversion of solar radiation into heat, absorbed by surfaces within the sun space, then reradiated out and back again.

Heat storage materials. Containers or materials used to store heat during the day, then release it slowly during the cooler night; water held in tubes, barrels or drums, masonry walls, or containers built into walls can all be used.

Low e glass. Coating applied to a glazing surface to reduce the amount of heat and light reaching the sun space.

Insolation. Radiation received from the sun.

Insulation. Material used to prevent absorption or loss of heat.

Laminated glass. Two layers of glass permanently bonded together with a layer of plastic material between them; acceptable for use in overhead applications, or in sliding doors.

Radiation. Energy emitted from the sun in the form of waves.

Tempered glass. Heat-treated glass that breaks into small, rounded fragments, not shards; acceptable for vertical, but not overhead, use.

Thermal breaks. Internal feature found in aluminum glazing bars to prevent conduction of cold from the outdoors to the interior of the sun space.

Accessories
Furnishings

Architectural Fiberglass
1330 Bellvue Street
P.O. Box 8100
Green Bay, WI 54308

Empire Garden West, Inc.
225 Sunrise Highway
Lynbrook, NY 11563

Erkins Studios, Inc.
604 Thames Street
Newport, RI 02840

The Florentine Craftsmen, Inc.
46-24 28th Street
Long Island City, NY 11101

Moultrie Manufacturing
P.O. Drawer 1179
Moultrie, GA 31776-1179

Reed Brothers
500 Turner Station
Sebastopol, CA 95472

Dan Wilson Company
Highway 401 North
P.O. Box 566
Fuquay-Varina, NC 27526

Gardening Catalogs

Charley's Greenhouse Supply
1569 Memorial Highway
Mount Vernon, WA 98273

Gardener's Eden
P.O. Box 7307
San Francisco, CA 94120

Smith & Hawken
25 Corte Madera
Mill Valley, CA 94941

Flooring

American Olean Tile Company
1000 Cannon Avenue
Lansdale, PA 19446

Bomanite Corporation
81 Encina Avenue
Palo Alto, CA 94301

Cal-Ga-Crete International, Inc.
803 Miraflores
San Pedro, CA 90731

Country Floors
15 East 16th Street
New York, NY 10003

Country Floors
8735 Melrose Avenue
Los Angeles, CA 90069

Ideal Tile
405 East 57th Street
New York, NY 10022

Shades and Screens

Appropriate Technology Corporation
P.O. Box 975
Brattleboro, VT 05301

Castec, Inc.
7531 Coldwater Canyon Boulevard
North Hollywood, CA 91605

Catalina Shading Systems
230 East Dyer Road, Unit G
Santa Ana, CA 92707

Comfortex Corporation
100 North Mohawk Street
P.O. Box 728
Cohoes, NY 12047

Dirt Road Company
R.D. 1, Box 260
Waitsfield, VT 05673

Shading Systems, Inc.
P.O. Box 5697
Clark, NJ 07066

Sol*R*Veil, Inc.
635 West 23rd Street
New York, NY 10011

Verosol USA, Inc.
215 Beecham Drive
P.O. Box 517
RIDC Park West
Pittsburgh, PA 15230

Vimco
9301 Old Staples Mill Road
Richmond, VA 23228

Engineers of Sun Space Designs

Steven Winter Associates, Inc.
6100 Empire State Building
New York, NY 10001

Steven Winter Associates, Inc.
170 Newport Center Drive
Newport Beach, CA 92660

Manufacturers of Sun Space Kits

Abundant Energy, Inc.
P.O. Box 307
Pine Island, NY 10969

Advance Energy Technologies, Inc.
P.O. Box 387
Clifton Park, NY 12065

Allstate Greenhouse Manufacturing Corporation
Box 89
Shoreham, NY 11786

American Solar Systems
13201 Hancock Drive
Taylor, MI 48180

Atria, Inc.
10301 North Enterprise Drive
Mequon, WI 53092

Better Products, Inc.
P.O. Box 1052
Alamosa, CO 81101

Brady & Sun, Inc.
97 Webster Street
Worcester, MA 01603

Brother Sun Glass and Window Center
2907 Agua Fria
Sante Fe, NM 87501

C-Thru Industries, Inc.
130 North Gilbert Street
Fullerton, CA 92633

California Solariums, Inc.
5300 North Irwindale Avenue
Irwindale, CA 91706

Classic Solar Design, Inc.
3 Steuben Drive
Jericho, NY 11753

Commonwealth Solaroom Greenhouse Corporation
Route 6, Box 302
Warrenton, VA 22186

Contemporary Structures, Inc.
1102 Center Street
Ludlow, MA 01056

Creative Structures, Inc.
1765 Walnut Lane
Quakertown, PA 18951

English Greenhouse Products Corporation
1501 Admiral Wilson Boulevard
Camden, NJ 08109

Evergreen Systems U.S.A., Inc.
P.O. Box 128
Burnett, WI 53922

Everlite Greenhouses, Inc.
9305-H Gerwig Lane
Columbia, MD 21046

Florian Greenhouse, Inc.
64 Airport Road
West Milford, NJ 07480

Four Seasons Solar Products Corporation
5005 Veterans Memorial Highway
Holbrook, NY 11741

Freedom Sunspace
R.D. 5, Box 172 Kinney Road
Freehold, NJ 07728

Gammans Industries, Inc.
P.O. Box 1181
Newnan, GA 30264

Garden Way Sunroom/Solar Greenhouse
430 Hudson River Road
Waterford, NY 12188

Glass Enclosures Corporation
6999 Hundtley Road, Suite N
Columbus, OH 43229

Gothic Arch Greenhouses
P.O. Box 1564
Mobile, AL 36633

Great Canadian Sunroom Company, Inc.
235 Lydia Street
Kitchener, Ontario N2H 1W4
Canada

Green Mountain Homes
Royalton, VT 05068

Habitek, Inc.
102 Queens Drive
King of Prussia, PA 19406

Holland House, Inc.
2012 Moore Street
Bellingham, WA 98226

Janco Greenhouses and Glass Structures
9390 Davis Avenue
Laurel, MD 20707

Keystone Shower Doors
P.O. Box 544
Southampton, PA 18966

LRC Products
P.O. Box 706
Warsaw, IN 46580

Lindal Cedar Homes, Inc.
4300 South 104th Place
Seattle, WA 98124

Lord & Burnham
Box 255
Irvington, NY 10533-0255

Machin Designs (USA), Inc.
557 Danbury Road (Route 7)
Wilton, CT 06897

National Greenhouse Company
P.O. Box 100
400 East Main Street
Pana, IL 62557

New England Glass Enclosures Inc.
50 Town Park Road #7
New Milford, CT 06776

Newcroft, Inc.
Box 464
Norwell, MA 02061

Northeast Sunspace, Inc.
799 (Route 3A) C.J. Cushing Way
Cohasset, MA 02025

Northern Sun, Inc.
15135 Northeast 90th Street, Building 5
Redmond, WA 98073

ODL, Inc.
215 East Roosevelt Avenue
Zeeland, MI 49464

Pacific Coast Greenhouse Manufacturing Company
8360 Industrial Avenue
Cotati, CA 94928

Paeco Industries, Inc.
P.O. Box 968,
1 Executive Drive
Toms River, NJ 08753

Pella Windows and Doors
102 Main Street
Pella, IA 50219

Proper Seal
1739 Savannah Highway
Charleston, SC 29407

Regal Manufacturing Company
P.O. Box 14578
Portland, OR 97214

Santa Barbara Greenhouses
1115 Avenue Acaso
Camarillo, CA 93010

Serres Solarium, Ltd.
1195 Rue Principale
Granby, Quebec J2G 868
Canada

Skytech Systems
P.O. Box 763
Bloomsburg, PA 17815

Solar Additions Sun Rooms
Route 50
Greenwich, NY 12834

Solar Components Corporation
88 Pine Street
Manchester, NH 03103

Solar Resources, Inc.
P.O. Box 1848
Taos, NM 87571

Solarium Systems, Inc.
P.O. Box 160
Mound, MN 55364

Solite Solar Greenhouses
1135 Bronx River Avenue
Bronx, NY 10472

Sunbilt Solar Products
109-10 180th Street
Jamaica, NY 11433

The Sun Company
14217 NE 200th Street
Woodinville, WA 98072

Sunglo Solar Greenhouses
4441 26th Avenue West
Seattle, WA 98199

Sunplace, Inc.
6601 Amberton Drive
Route 100 Business Park
Elkridge, MD 21227

Sun Room Company, Inc.
322 East Main Street
Leola, PA 17540

Sun Room Designs, Inc.
Depot & First Streets
Youngwood, PA 15697

Sunshine Rooms, Inc.
P.O. Box 4627
Wichita, KS 67204

Sun System Solar Greenhouses
60 Vanderbilt Motor Parkway
Commack, NY 11725

Sunworks Greenhouses
3060 South 24th Street
Kansas City, KS 66106

Supreme Solariums
P.O. Box 3765
Eugene, OR 97403

Texas Greenhouse Company
2717 St. Louis Avenue
Fort Worth, TX 76110

Turner Greenhouses
Highway 117 Smith
P.O. Box 1260
Goldsboro, NC 27530

Westview Products, Inc.
P.O. Box 569
Dallas, OR 97338

Vegetable Factory, Inc.
71 Vanderbilt
New York, NY 10169

Organizations

American Concrete Institute
P.O. Box 19150
Redford Station
Detroit, MI 48219

Brick Institute of America
11490 Commerce Park Drive
Suite 900
Reston, VA 22091

Laminators Safety Glass Association
3310 Harrison
Topeka, KS 66611

Marble Institute of America
33505 State Street
Farmington, MI 48024

National Concrete Masonry Association
2302 Horse Pen Road
Herndon, VA 22070

National Greenhouse Manufacturers Association
P.O. Box 567
Pana, IL 62557

National Window Fashions Association
372 St. Peter Street, Suite 310
St. Paul, MN 55102

Tile Council of America, Inc.
P.O. Box 2222
Princeton, NJ 08540